Wheelchair
Skills Assessment
and **Training**

REHABILITATION SCIENCE IN PRACTICE SERIES

Series Editors

Marcia J. Scherer, Ph.D.

President
*Institute for Matching Person and
Technology*

Professor
*Physical Medicine & Rehabilitation
University of Rochester Medical Center*

Dave Muller, Ph.D.

Executive
Suffolk New College

Editor-in-Chief
Disability and Rehabilitation

Founding Editor
Aphasiology

Published Titles

Ambient Assisted Living, *Nuno M. Garcia and Joel J.P.C. Rodrigues*

Assistive Technology Assessment Handbook, *edited by Stefano Federici and Marcia J. Scherer*

Assistive Technology for Blindness and Low Vision, *Roberto Manduchi and Sri Kurniawan*

Computer Access for People with Disabilities: A Human Factors Approach,
    Richard C. Simpson

Computer Systems Experiences of Users with and Without Disabilities: An Evaluation Guide
    for Professionals, *Simone Borsci, Maria Laura Mele, Masaaki Kurosu, and Stefano Federici*

Devices for Mobility and Manipulation for People with Reduced Abilities,
    Teodiano Bastos-Filho, Dinesh Kumar, and Sridhar Poosapadi Arjunan

Human–Computer Interface Technologies for the Motor Impaired,
    edited by Dinesh K. Kumar and Sridhar Poosapadi Arjunan

Multiple Sclerosis Rehabilitation: From Impairment to Participation,
    edited by Marcia Finlayson

Neuroprosthetics: Principles and Applications, *edited by Justin Sanchez*

Paediatric Rehabilitation Engineering: From Disability to Possibility, *edited by Tom Chau
    and Jillian Fairley*

Quality of Life Technology Handbook, *Richard Schultz*

Rehabilitation: A Post-critical Approach, *Barbara E. Gibson*

Rehabilitation Goal Setting: Theory, Practice and Evidence, *edited by Richard J. Siegert
    and William M. M. Levack*

Rethinking Rehabilitation: Theory and Practice, *edited by Kathryn McPherson,
    Barbara E. Gibson, and Alain Leplège*

Wheelchair Skills Assessment and Training, *R. Lee Kirby*

Wheelchair Skills Assessment and Training

R. Lee Kirby

Dalhousie University, Halifax, Canada

CRC Press

Taylor & Francis Group

Boca Raton   London   New York

CRC Press is an imprint of the
Taylor & Francis Group, an **informa** business

CRC Press
Taylor & Francis Group
6000 Broken Sound Parkway NW, Suite 300
Boca Raton, FL 33487-2742

First issued in paperback 2021

© 2017 by Tartan Rehab Limited
CRC Press is an imprint of Taylor & Francis Group, an Informa business

No claim to original U.S. Government works

ISBN 13: 978-1-4987-3881-1 (hbk)
ISBN 13: 978-1-03-224044-2 (pbk)

DOI: 10.1201/9781315369389

Library of Congress Cataloging-in-Publication Data

Names: Kirby, R. Lee, author.
Title: Wheelchair skills assessment and training / R. Lee Kirby.
Other titles: Rehabilitation science in practice series. 2469-5513
Description: Boca Raton : Taylor & Francis, 2017. | Series: Rehabilitation science in practice series | Includes bibliographical references and index.
Identifiers: LCCN 2016032442 | ISBN 9781498738811 (hardcover : alk. paper)
Subjects: | MESH: Wheelchairs | Patient Education as Topic--methods | Disabled Persons--rehabilitation | Motor Skills | Disability Evaluation | Outcome Assessment (Health Care)--methods
Classification: LCC RD757.W4 | NLM WB 320 | DDC 617/.033--dc23
LC record available at https://lccn.loc.gov/2016032442

Visit the Taylor & Francis Web site at
http://www.taylorandfrancis.com

and the CRC Press Web site at
http://www.crcpress.com

Publisher's Note
The publisher has gone to great lengths to ensure the quality of this reprint but points out that some imperfections in the original copies may be apparent.

To

Patty and the Wheelchair Research Team

For

*People who use wheelchairs, their caregivers,
and their healthcare workers*

Contents

Warnings, disclaimers, and conditions of use

Some of the wheelchair skills addressed in this book can be dangerous and may result in severe injury or death if attempted without the assistance of trained personnel. Attempting these skills may not be appropriate for some wheelchair users or caregivers. If the skills are attempted for assessment or training purposes, one or more experienced spotters should be available to intervene. The ultimate responsibility for safety during the performance of wheelchair skills lies with the person performing the skills. Even if a spotter acts according to instructions, injuries can occur. Safely performing a skill in the supervised Wheelchair Skills Program (WSP) context provides no guarantee that the same or similar skills will be performed safely on other occasions.

The information in this book is provided for information and educational purposes only. The information is not intended to be and does not constitute healthcare advice. Any decision concerning the health, treatment, and/or wheelchair of a wheelchair user or caregiver should be made in consultation with a qualified healthcare professional. The members of the team who developed the WSP, the members of the Editorial Committee, Dalhousie University, and the Nova Scotia Health Authority are not responsible for any injuries or deaths arising from the use of the WSP materials. Users of these materials assume full responsibility for their actions.

Anyone wishing to use the WSP materials may do so without permission or charge, as long as they accept and comply with the Conditions of Use posted on the WSP website (http://www.wheelchairskillsprogram.ca/eng/conditions.php).

Acknowledgments

This book is based on the Wheelchair Skills Program (WSP) 4.3 Manual (http://www.wheelchairskillsprogram.ca/eng/manual.php) and related materials. These materials are the result of the work of many people. Those who have had the greatest recent involvement constitute the members of the Editorial Committee, namely, Cher Smith, BScOT, MSc; Kim Parker, MASc, PEng; Mike McAllister, PhD; Joy Boyce, Hons, BScOT; Paula W. Rushton, PhD; François Routhier, PhD; Krista L. Best, PhD; Diane MacKenzie, PhD; Ben Mortenson, PhD; and Åse Brandt, PhD. I thank Danny Abriel of Design Services at Dalhousie University and the models for the photographs that illustrate the book.

The list of colleagues who have contributed and research funding agencies that have supported this work is extensive and continues to grow. Their names are noted in specific published articles, which can be found in the references listed in Chapter 9 of this book and on the WSP website (www.wheelchairskillsprogram.ca/eng/publications.php).

Similarly, this work would not have been possible without the many excellent papers, textbooks, and training manuals that have been published by others. Some of this literature has been acknowledged in the reference sections of the published articles in Chapter 9.

I also thank those at CRC Press and Taylor and Francis, in particular Michael Slaughter, S.K. Prasanna and Ed Curtis, for all their hard work in making this project a reality.

Author

 **R. Lee Kirby** earned an MD at Dalhousie University in Halifax, Nova Scotia, Canada. His specialty training in physical medicine and rehabilitation was carried out at the University of Washington in Seattle, Washington, at Dalhousie University in Halifax, and at Stoke Mandeville Hospital in England. He is a professor in the Division of Physical Medicine and Rehabilitation in the Department of Medicine at Dalhousie University in Halifax with a cross-appointment in Community Health and Epidemiology. He is based at the Nova Scotia Rehabilitation and Arthritis Centre Site of the Nova Scotia Health Authority. His primary research interest is the safety and performance of wheelchairs. He has held research grants from a number of national and international funding bodies. He has published numerous articles in peer-reviewed journals and made many presentations to national or international meetings. He heads the team that developed the Wheelchair Skills Program.

List of abbreviations

CG	Caregiver
GAS	Goal Attainment Score
NP	Not possible
RAD	Rear anti-tip device
SCI	Spinal cord injury
SU	Scooter user
TE	Testing error
UN	United Nations
WC	Wheelchair
WCU	Wheelchair user
WHO	World Health Organization
WSP	Wheelchair Skills Program
WST	Wheelchair Skills Test
WSTP	Wheelchair Skills Training Program
WST-Q	Questionnaire version of the WST

Prologue

In the 2008 World Health Organization (WHO) Guidelines on the Provision of Wheelchairs in Less-Resourced Settings (www.who.int/disabilities/publications/technology/wheelchairguidelines/en/), it was estimated there were 65 million people globally who would benefit from wheelchairs but that 20 million of these people did not have access to them. The prevalence of wheelchair use is rising, in part due to the aging of the population. Of the wheelchairs in use in highly developed parts of the world, about 70% are manual wheelchairs, with the remainder divided about equally between powered wheelchairs and scooters.

The wheelchair is arguably the most important therapeutic tool in rehabilitation. Research studies have documented such benefits as improved mobility, improved participation, reduced caregiver burden, and reduced likelihood of placement in long-term care facilities. Yet, despite the importance of wheelchairs, they are far from perfect. Many wheelchairs are inappropriate for their users, fit them poorly, or are poorly set up. Repairs are needed often and many wheelchair users suffer from acute or chronic injuries due to wheelchair use. Improvements in safety often come at the expense of performance and vice versa. For instance, a highly stable manual wheelchair may be less likely to tip over but will create problems when the wheelchair user attempts to unload the front wheels (casters) to overcome obstacles. Inaccessibility restricts the usefulness of wheelchairs for some users.

The manner in which people receive wheelchairs varies widely. At the "commodity" end of the spectrum, a wheelchair can be purchased without any clinical input, "over the counter" at the corner drugstore. Optimally, as described by the 2008 WHO Guidelines on the Provision of Wheelchairs in Less Resourced Settings, there is a care pathway that includes assessment by professionals, the development of a prescription with the involvement of the wheelchair user and family, assistance (if needed) with the organization of funding for the wheelchair, proper fitting and adjustment of the wheelchair, training of the wheelchair user and caregiver in maintenance and handling skills, and long-term follow-up for refinements, routine servicing, and periodic replacement.

Two important elements in this care pathway are wheelchair skills assessment and training for wheelchair users and their caregivers. The WSP is a set of assessment and training protocols related to wheelchair skills. Wheelchair skills assessment and training are topics that have received relatively little attention until the past two decades. What has spurred current interest in this topic has been an accumulating body of research evidence.

Skill in wheelchair use is not an end in itself; it is a means to an end. In terms of the WHO's International Classification of Function (2001), wheelchair skills are "activities." The ability to perform them represents "capacity" and their use in everyday life represents "performance." The purpose of these activities is to overcome barriers in the environment and to thereby permit the wheelchair user to fulfil his/her desired role in society ("participation"). Other potential benefits of wheelchair-skills training for wheelchair users and caregivers include fewer acute and overuse injuries, an improved sense of wellbeing (through self-esteem, confidence and personal control, the sense of becoming newly enabled or empowered, and having accomplished something of worth), improved development (of children), and having fun.

In addition to or instead of learning wheelchair skills, there may be alternative ways to accomplish the learner's goals (e.g., by changing wheelchairs, by accepting the assistance of a caregiver, by eliminating accessibility barriers). Alternatively, if the goal of performing a wheelchair skill proves not to be a feasible one, the most appropriate strategy may be to assist the learner in adjusting his/her expectations to a more realistic level.

Although there are many similarities in how to best perform a skill, regardless of the characteristics of the wheelchair user and the impairments that have led to wheelchair use, there are also differences. What is safe and effective for a young fit woman with incomplete paraplegia may be different for a middle-aged overweight man with complete tetraplegia, and even more different for the elderly foot-propelling person with a stroke.

The characteristics of the wheelchair—its features, fit, and setup—can have major effects on skill performance. In helping improve the safety, effectiveness, and efficiency of wheelchair use, service-delivery providers should try to optimize the wheelchair user (e.g., by improving strength or range of motion), the wheelchair (e.g., by moving the axles of a manual wheelchair forward, adjusting the programming of a powered wheelchair), and/or training.

Major independent bodies such as the United Nations (UN) (Convention on the Rights of Persons with Disabilities, 2006) and the WHO (Guidelines on the Provision of Wheelchairs in Less-Resourced Settings, 2008) have endorsed the importance of skills training.

The Wheelchair Research Team at Dalhousie University and the Nova Scotia Rehabilitation Centre in Halifax, Nova Scotia, Canada began in the early 1980s with a research project to determine why rehabilitation professionals were observing that recently developed lightweight wheelchairs were tipping over as often as they were. This was followed by a series of research studies that developed testing methods and answered questions about the nature of static and dynamic stability of occupied wheelchairs.

The work on dynamic stability led to the development of the Wheelchair Skills Test (WST) in 1996 as a means of assessing the ability of wheelchair users to safely perform the skills they needed in their everyday lives. Subsequently, a questionnaire version (the WST-Q) has been added. There have been a growing number of peer-reviewed articles (www.wheelchairskillsprogram.ca/eng/publications.php) (see Chapter 9.1) about the measurement properties of the WST/WST-Q or that have used the WST/WST-Q as an outcome measure in other studies.

Having developed a useful measurement tool, it became apparent that many wheelchair users could not perform all of the skills that might be helpful to them. This led to the development of the Wheelchair Skills Training Program (WSTP), using the best-available evidence on motor skills learning principles and the best-available evidence on wheelchair skill techniques. Since then, there have been a growing number of peer-reviewed articles (www.wheelchairskillsprogram.ca/eng/publications.php) (see Chapter 9.1) that have documented the safety and effectiveness of such training.

The WSP is a set of protocols for the assessment and training of wheelchair skills— the WST/WST-Q and WSTP, respectively. The WSP has expanded its scope from manual wheelchairs to include powered wheelchairs and scooters and to include caregivers in addition to wheelchair users. The WSP website, all of the materials on which are provided free of charge, had 113,278 visits from 72,971 users in 177 countries as of September 19, 2016. Members of the Wheelchair Research Team have provided practical training on the WSP to therapists in a number of countries around the world, in both highly developed and less-resourced settings. The WSP is now recognized by a variety of national and international organizations.

The WSP has evolved over time, in response to feedback and experience with it. Various iterations of the WSP—#1.0, 2.4, 3.2, 4.1, 4.2, and 4.3 to date—have been released for general use. WSP 4.3 (i.e., the basis for this book) was originally released for use on November 6, 2015. Even within the lifespan of an iteration, the WSP materials are periodically updated. As such, the materials are "living" rather than fixed. If the iteration number has not changed (e.g., from 4.2 to 4.3) despite an update, it is because the changes have been deemed by the Editorial Committee to be predominantly of a minor nature. However, for academic purposes, users of the WSP materials should cite the date of the iteration that they use. This can be found in the footer of each page.

WSP 4.3 is different from WSP 4.2 in the following notable ways:

- The skill set has evolved slightly.
- Some of the skills have been renamed to make them easier to understand.
- The order of the skills has been revised to better reflect their relative difficulty.
- Confidence scoring has been added to the WST-Q.

The WSP is different from most other resources on wheelchair skills in a number of ways:

- It is based on the best evidence on how to perform, assess, and teach wheelchair skills.
- Where there are gaps in evidence, ongoing evaluation of the WSP has been initiated with as much scientific rigor as possible.
- The process and sequence of the training has evolved.
- The materials are continuously being updated.
- The WSP deals with both assessment and training.
- The WSP deals with the skills of the wheelchair users themselves, alone or in combination with their caregivers.
- The WSP deals with the full spectrum of wheelchair users (e.g., hand propellers such as those using wheelchairs due to spinal cord injury (SCI) as well as foot propellers such as those using wheelchairs due to stroke or dementia).
- The WSP deals with manual wheelchairs, powered wheelchairs, and scooters.
- All of the materials on the WSP website have been made available free of charge ("open source").

In this book, I have attempted to provide a wide spectrum of readers with comprehensive but easily understandable materials. The target audience includes practicing and student rehabilitation therapists (e.g., occupational, physical, recreational), their aids and assistants, rehabilitation nurses and rehabilitation medicine physicians and residents. In addition to clinicians, researchers and their staff may find the book to be a useful resource. Additionally, because the book has been written in plain language, many wheelchair users and caregivers should be able to comprehend the contents with ease. Because the assessment and training of wheelchair skills are low-tech and the training program is high-impact, the WSP is equally relevant for highly developed and less-resourced parts of the world.

As recommended in the WHO Guidelines, a new wheelchair user should go through an eight-step process in the course of his/her wheelchair service delivery. One of those steps is assessment. As part of this assessment, the wheelchair skills of the wheelchair user should be assessed. This should be done at intake, as part of the prescription and fitting steps (e.g., to compare how well the wheelchair user can perform skills with a rigid vs. a folding wheelchair, or with the rear axles in more and less stable positions) and during follow-up to determine the revisions required in the wheelchair. The assessment can be performed using the WST or the WST-Q. Another WHO step is training, which includes wheelchair skills training of the wheelchair user and/or caregiver. For this training, the WSTP can be used during the initial provision of the wheelchair and as necessary at follow-up. The WHO's eight steps of wheelchair service delivery need not be sequential and are often iterative. For instance, following training, it may be possible to revise the prescription and set-up.

This book is comprised of chapters dealing with overviews of the WSP, the assessment of wheelchair skills, training, and safety issues. Following that, each of the individual skills that make up the WSP skill set is dealt with in detail. Near the end of the book, there is a chapter on games that can be played to reinforce learning. A chapter on suggested reading provides a starting point for those interested in further study. Finally, appendices provide some options regarding the organization of training sessions for individuals or groups as well as some of the forms that can be used.

Chapter 1 Introduction to the Wheelchair Skills Program (WSP)

1.1 Scope

The WSP is intended for manual wheelchairs (Figure 1.1) or powered wheelchairs (Figure 1.2), operated by wheelchair users or caregivers. The WSP can also be used for scooter users (Figure 1.3). Whenever appropriate, the word "wheelchair" used in the book should be understood to include scooters. Throughout the book, to simplify descriptions, unless otherwise specified, it has been assumed that the wheelchair being used is one with rear-wheel drive (i.e., large diameter wheels in back and smaller diameter swivel or steerable casters in front). Other types of wheelchairs and scooters can be dealt with using the WSP materials, but some of the instructions and explanations may need to be adapted accordingly. Wheelchair technology is diverse and evolving at a rapid rate. There may be wheelchairs that do not easily fit the categories described. In such situations, the tester or trainer needs to exercise judgment regarding which skills are appropriate. For instance, for power-assisted wheelchairs, a combination of skills from the manual and powered wheelchair skill sets would be appropriate.

The WSP is not intended to be an adequate approach for other important wheelchair skills (e.g., maintenance and repair skills), more extreme skills (e.g., some wheelchair sport activities), or community-integration activities that combine a number of skills (e.g., use of accessible transport and shopping). The skills chosen for inclusion in the WSP are intended to be representative of the range of skills that wheelchair users and/or caregivers may need to regularly perform, varying from the most basic to the more difficult. However, it would be impossible to be all-inclusive without making the size of the WSP unmanageable.

1.2 Subjects

In the book, the term "subject" is often used because the person who is the object of testing or training may be a wheelchair user, a caregiver, a healthcare student, or a research participant. In addition to testing or training for a wheelchair user

1

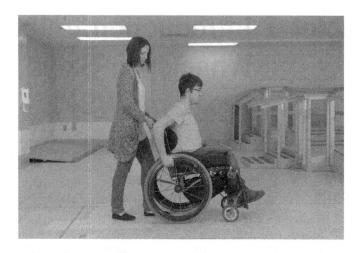

Figure 1.1 Manual wheelchair.

and a single caregiver separately, the WSP may be used to assess or train the extent to which one or more caregivers and a wheelchair user can function as a team; the "subject" in such situations is the combination of the wheelchair user and the caregiver(s). Unless otherwise specified, the assumption is that it is a single person operating alone who is the subject. If an animal (e.g., a service dog) is used to assist with a skill, the animal is considered an "aid" rather than a caregiver.

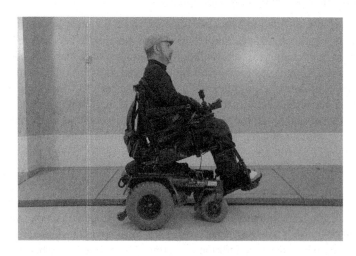

Figure 1.2 Powered wheelchair with rear-wheel drive.

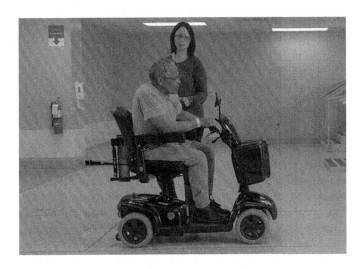

Figure 1.3 Scooter.

1.3 Special considerations for caregivers

If the usual circumstance for a skill in real life is that a wheelchair user and his/her caregiver ordinarily share the duties, then "blended" wheelchair user/caregiver testing or training may be the most appropriate choice, but the relative roles of the two people involved should be noted. It is not a reasonable expectation that a single caregiver could complete some skills alone without special equipment or the cooperation of the wheelchair user. The test score achieved is often a combination of the capacity of the wheelchair user and caregiver functioning together (i.e., a "blended" score). The caregiver-assisted score is specific not only to the wheelchair and setting (as is the case for the WST in general), but is also specific to the wheelchair user being assisted.

In addition to physical assistance, caregiver assistance may be just the presence of the caregiver (for reassurance, moral support, and spotting) without necessarily any cues being provided (i.e., "standby assistance"). If a caregiver is the subject of testing, he/she must meet the same criteria used for the wheelchair user (e.g., keeping the caregiver's feet as well as the wheelchair wheels inside any designated limits). Special additional caregiver considerations are noted in the later chapter on individual skills.

1.4 The circle of education

Assessment and training are both elements in the classical circle of education. In this circle, one begins with an assessment (including the WST or

3

WST-Q) to identify the learner's starting point. From this, the educational objectives are individualized. This is followed by the curriculum (the training program), aimed at meeting these objectives. This is followed by another assessment to confirm that the objectives have been met. If not, the cycle continues.

1.5 Cost-effectiveness of the WSP

Although no formal studies of cost-effectiveness have yet been conducted, there is some basis for believing the WSP to be highly cost-effective. The WST requires an average of about 30 minutes to conduct and the WST-Q about 10 minutes. The training studies to date suggest that improvements in capability can be accomplished with as little as 4 hours of training (although many more are recommended). No equipment is required, only trained personnel. For personnel, occupational therapists, physical therapists and their assistants have the most appropriate backgrounds. However, there have also been good results when using university students or research assistants as testers and trainers. Learning a new skill lasts a lifetime, unlike strength or endurance training that requires ongoing efforts to maintain benefits. For all of these reasons, the WSP can be considered to be a cost-effective intervention that would compare favorably with other rehabilitation assessment measures or interventions.

1.6 Languages

The WSP was originally developed in the English language. It has since been translated by a team led by Francois Routhier (a member of the Wheelchair Skills Program Editorial Committee) into French (http://www.wheelchairskillsprogram. ca/fr/). Translation into other languages is encouraged and there have been some initiatives in other countries (to some of which the WSP website provides links—http://www.wheelchairskillsprogram.ca/eng/links.php).

1.7 Initial interview

Wheelchair skills assessment and training in the clinical setting usually takes place as part of a broader process related to the wheelchair user's and/or caregiver's health, function, and context. Prior to beginning testing or training, the tester or trainer should screen the subject for the ability to communicate and should obtain consent to proceed. If appropriate, demographic, clinical, and wheelchair-related data are recorded. These data may be obtained from the wheelchair user, the caregiver, and/or the health record.

1.8 Wheelchair and subject set-up

The wheelchair user and/or caregiver should be dressed and equipped as usual (e.g., wearing artificial limbs, braces) when using the wheelchair. The wheelchair should be set up as usual for that user. This is important because changes in the personal equipment or wheelchair set-up can affect how and how well the skills are performed.

If the wheelchair has user-adjustable features that could affect handling (e.g., rear anti-tip devices [RADs] for a manual wheelchair, a more powerful controller mode for a powered wheelchair), the subject is permitted to adjust them for testing as long as the subject can do so without assistance. If tools are needed to make the adjustment, then they must be carried by the subject. The tester must not cue the test subject to make the adjustment. Having adjusted the wheelchair to accomplish a skill, unless otherwise specified, the subject may leave the wheelchair in the new configuration for the remainder of the WST. If the subject wishes to restore the wheelchair to its original configuration, he/she must do so without assistance and without cueing from the tester. When the WST is over, the tester should remind the subject about any adjustment that has been made, especially if the adjustment might affect safety.

For training purposes, unless otherwise specified for the purposes of a research study, the wheelchair set-up may and should be modified if such a change is believed by the trainer to be one that would improve the safety or effectiveness of the skill performance.

1.9 Getting out of the wheelchair to accomplish a task

If possible to do so safely, a wheelchair user may get out of the wheelchair to accomplish a task or to adjust a wheelchair feature (e.g., the RADs). For the WST, this does not include using any sitting surface other than the ground, unless specifically noted in the individual skill chapter, because such a surface might not always be available when such an adjustment is needed. The policy of permitting wheelchair users to get out of their wheelchairs is in recognition that many people who use wheelchairs do so in combination with walking for their mobility.

1.10 Starting positions

Unless otherwise noted, the starting positions for each skill are as follows:

Wheelchair user: The wheelchair user is seated in the wheelchair, in whatever position and state that he/she prefers.

Caregiver: If a caregiver is the subject, his/her starting position is generally standing near the wheelchair.

Wheelchair: All of the wheelchair components that are usually used should be in place. The wheel locks (brakes) may be locked or unlocked. A rolling start is permitted (i.e., there is no need to come to a complete stop before beginning the skill attempt). When a starting position for the wheelchair is defined (e.g., relative to an obstacle), the tester may assist the subject in getting into this position. The tester should be careful not to provide inadvertent cues to the subject on how to perform the skill. For instance, with a powered wheelchair that has both caregiver- and user-operated controls, the tester should use the caregiver controls because they are usually out of the wheelchair user's line of sight. If the subject expresses the wish to attempt a task by moving the wheelchair backward, the tester may assist him/her in getting into the requested starting position, but the tester must not suggest alternative approaches. Also, when the testing instructions call for the axles of the leading wheels to be behind a starting line, the leading wheels are ones that are normally in contact with the ground (i.e., not the wheels of anti-tip devices that are off the ground).

Tester or trainer: The starting position for the tester or trainer is initially where he/she can be well seen and heard when providing instructions for the skill. After initially communicating instructions to the subject, the tester or trainer may need to reposition himself/herself where he/she will be best able to observe the skill.

Spotter: The starting position for the spotter is near the wheelchair (within an arm's reach) where he/she will be best able to respond to any safety concerns. The exact position varies with the skill being attempted, the number of spotters involved and the method being used to complete the skill. For powered wheelchairs, the spotter should be in a position where the power can be turned off or the joystick accessed. If a caregiver is the subject, he/she is expected to behave in a manner that is safe for both the wheelchair occupant and the caregiver. The spotter in such situations should remain close enough to intervene if the caregiver fails to exercise due caution.

If the alternative starting positions are different for a specific skill, the starting positions are specified in the chapter on individual skills.

1.11 Warnings to subject

Prior to beginning the initial WSP session, the subject should be warned by the WSP personnel that some wheelchair skills can be dangerous and that the subject or learner should not attempt any task that he/she is not comfortable performing. Also, to avoid overuse injury, the subject should be instructed to avoid overexerting himself/herself in the mistaken belief that success on every skill is expected.

These warnings may be repeated at any time during a WSP session. If, with the subject's knowledge and permission, the RADs are adjusted or removed, the WSP personnel should inform the subject that this has been done.

1.12 WSP personnel

WSP personnel are important elements in testing and training. During WSP activities, the roles of the tester and trainer are primarily to oversee the assessment and training of participants. The spotter is the person, other than the person performing the skill, who is primarily responsible for ensuring the safety of the subject from the moment the session begins until it is completed. The spotter focuses on the prevention of major acute injury. Wheelchair users, caregivers, testers, and trainers also play a role in preventing injury.

Although it is common for the tester or trainer to simultaneously fulfill the role of the spotter, it is useful to consider the roles separately. Although related, the competencies of spotters are different from those of testers and trainers. If the spotter and tester or trainer roles are being fulfilled by different people, and there is a difference of opinion between the WSP personnel, the tester or trainer shall make the final decision, after carefully considering the opinion of the spotter. Ordinarily, a single spotter can adequately minimize the likelihood of serious injury.

However, for some situations (e.g., a heavy or impulsive wheelchair user), one or more additional spotters may be needed. If more than one spotter is used, one spotter should take the lead role. Although testers and trainers need not be able to perform the physical spotter tasks themselves, they should understand the spotter's role and be able to supervise the spotter.

WSP personnel may be rehabilitation clinicians who are regularly involved in wheelchair provision, but there are no minimal educational prerequisites. However, WSP personnel should be thoroughly familiar with all elements of the WSP for which they have responsibility. WSP personnel should feel free to refer to the book whenever necessary.

Those interested in becoming WSP personnel should read this book, study related materials, review practice materials (e.g., videos on the website) and, if possible, observe in-person how experienced WSP personnel function. Ideally, the WSP should only be used by personnel who have been trained in its administration. However, good results are possible by careful attention to the book because the materials have been designed to be reasonably self-explanatory and to reflect normal clinical practices.

Because practice outside formal training sessions can be useful, members of the rehabilitation team (e.g., members of the nursing profession, personal care workers, recreation therapists, volunteers, physicians) other than the primary trainer can be of assistance. Good team communication among team members about a learner's progress can help to ensure that the input from multiple team members is complementary rather than conflicting. Because the principles of motor skills learning used for wheelchair skills are the same as those used when learning other skills (e.g., music, sport), a background in teaching such other motor skills is an asset for a trainer. Similarly, experience in managing groups (e.g., coaching sports, supervising children) is an asset to any trainer teaching wheelchair skills in a group setting.

Both experts and nonexperts can play important roles in the training process. Wheelchair-using or caregiver peers may possess or be able to acquire the necessary knowledge, skills, and attitudes to function as WSP personnel. Peers have a number of advantages over able-bodied personnel—real-life experience with barriers, familiarity with practical solutions to common problems, credibility, and superior capacity to empathize with the difficulties being experienced by a wheelchair-using subject. However, the peer may have limited clinical knowledge (e.g., about what triggers a spasm), his/her expertise in performing wheelchair skills is likely to be highly specific (e.g., a peer with a SCI may have difficulty advising a person using a wheelchair due to a stroke) and a wheelchair user may have difficulty functioning as a spotter for some skills (particularly moving skills).

The personal characteristics of WSP personnel are also important. Personnel should be credible, friendly, supportive, non-judgmental, interested, and honest. Personnel should be familiar with the structure and operation of the specific wheelchair used by the subject.

1.13 Versions of the WSP

There are five modular versions of the WSP (Table 1.1). The version selected for use is based on the type of wheelchair and the nature of the subject or learner.

Table 1.1 Versions of the WSP by Type of Wheelchair and Nature of the Subject

Version #	Type of Wheelchair	Type of Subject
1	Manual	Wheelchair user
2	Manual	Caregiver
3	Powered	Wheelchair user
4	Powered	Caregiver
5	Scooter	Scooter user

1.14 Individual skills

The individual skills (Table 1.2) are the units of assessment and training. These skills can be put together in various combinations and permutations to allow participation (e.g., going shopping, attending an educational event, performing a job). A brief description of each skill and the rationale for including it in the WSP can be found in the later chapter on individual skills. In naming the individual skills, the attempt has been to be as generic and universal as possible. This is in recognition that the environments in which wheelchairs are used vary widely around the world, although they share many common characteristics.

The WST, the WST-Q, and the WSTP all deal with the same set of skills, but the correspondence should not be considered exact. For instance, for the "rolls forward short distance" skill, the WST by necessity deals with exact dimensions (10 m), the WST-Q questions are stated in more general terms because subjects may not be able to easily visualize such exact distances and the WSTP involves variations that enhance learning. The order of skills in Table 1.2 reflects the functional groupings of skills (e.g., inclines in different directions and with different slopes are grouped together) and the approximate order of difficulty (although this can vary depending upon the subject and wheelchair).

1.15 Skill levels

Although somewhat arbitrary, it is possible on the basis of difficulty to roughly group skills into three levels—basic, intermediate, and advanced. This can be helpful for communicating with others, for planning therapies, and for justifying the purchase of different types of wheelchairs. Which skills have been assigned to which levels is indicated in Table 1.2. If appropriate to the circumstances, users of the WSP may select only skills of one level to focus on (e.g., the basic level for residents of a long-term-care facility who rarely use their wheelchairs in the community without assistance).

1.16 Skill groups

Most of the individual skills can be grouped, as described below, although some of these groupings only apply to manual wheelchairs.

How to operate the parts of the wheelchair: Wheelchairs vary widely in their components and how they work. It is important that wheelchair users and caregivers learn about the structures and operating idiosyncrasies of the wheelchairs they use. This includes normal daily operations, transportation, and storage of the wheelchair, as well as regular

Table 1.2 List of Individual Skills in Wheelchair Skills Program 4.3

#	Skill Level	Individual Skill Names	Manual WC		Powered WC		Scooter
			WCU	CG	WCU	CG	SU
1.	Basic	Moves controller away and back	✗	✗	✓	✓	✓
2.	Basic	Turns power on and off	✗	✗	✓	✓	✓
3.	Intermediate	Selects drive modes and speeds	✗	✗	✓	✓	✓
4.	Basic	Disengages and engages motors	✗	✗	✓	✓	✓
5.	Basic	Operates battery charger	✗	✗	✓	✓	✓
6.	Basic	Rolls forward short distance	✓	✓	✓	✓	✓
7.	Basic	Rolls backward short distance	✓	✓	✓	✓	✓
8.	Basic	Turns in place	✓	✓	✓	✓	✓
9.	Basic	Turns while moving forward	✓	✓	✓	✓	✓
10.	Basic	Turns while moving backward	✓	✓	✓	✓	✓
11.	Basic	Maneuvers sideways	✓	✓	✓	✓	✓
12.	Basic	Reaches high object	✓	✗	✓	✗	✓
13.	Basic	Picks object from floor	✓	✓	✓	✓	✓
14.	Basic	Relieves weight from buttocks	✓	✓	✓	✓	✗
15.	Basic	Operates body positioning options	✓	✓	✓	✓	✓
16.	Basic	Level transfer	✓	✓	✓	✓	✓
17.	Intermediate	Folds and unfolds wheelchair	✓	✓	✗	✗	✗
18.	Intermediate	Gets through hinged door	✓	✓	✓	✓	✓
19.	Intermediate	Rolls longer distance	✓	✓	✓	✓	✓
20.	Intermediate	Avoids moving obstacles	✓	✓	✓	✓	✓

(Continued)

10

Table 1.2 (*Continued*) List of Individual Skills in Wheelchair Skills Program 4.3

#	Skill Level	Individual Skill Names	Manual WC		Powered WC		Scooter
			WCU	CG	WCU	CG	SU
21.	Intermediate	Ascends slight incline	✓	✓	✓	✓	✓
22.	Intermediate	Descends slight incline	◆✓	✓	✓	✓	✓
23.	Advanced	Ascends steep incline	✓	✓	✓	✓	✓
24.	Advanced	Descends steep incline	✓	✓	✓	✓	✓
25.	Intermediate	Rolls across side-slope	✓	✓	✓	✓	✓
26.	Intermediate	Rolls on soft surface	✓	✓	✓	✓	✓
27.	Intermediate	Gets over threshold	✓	✓	✓	✓	✓
28.	Intermediate	Gets over gap	✓	✓	✓	✓	✓
29.	Intermediate	Ascends low curb	✓	✓	✓	✓	✓
30.	Intermediate	Descends low curb	✓	✓	✓	✓	✓
31.	Advanced	Ascends high curb	✓	✓	✗	✗	✗
32.	Advanced	Descends high curb	✓	✓	✗	✗	✗
33.	Advanced	Performs stationary wheelie	✓	✓	✗	✗	✗
34.	Advanced	Turns in place in wheelie position	✓	✓	✗	✗	✗
35.	Advanced	Descends high curb in wheelie position	✓	✓	✗	✗	✗
36.	Advanced	Descends steep incline in wheelie position	✓	✓	✗	✗	✗
37.	Advanced	Gets from ground into wheelchair	✓	✓	✓	✓	✓
38.	Advanced	Ascends stairs	✓	✓	✗	✗	✗
39.	Advanced	Descends stairs	✓	✓	✗	✗	✗

Note: Abbreviations and symbols: WC = wheelchair, WCU = wheelchair user, CG = caregiver, SU = scooter user, ✓ = included, ✗ = not included.

maintenance duties. At the time of sale, new wheelchairs are usually delivered with user manuals. Wheelchair users and caregivers can learn about special features of their wheelchairs by studying the user manuals. If the user manual has been lost, instructions can often be found online. Maintenance and repair issues are also usually described in the user manual (e.g., how to recognize when maintenance or repair are needed, how often a battery needs to be charged).

Understanding the dimensions of the wheelchair: The dimensions of the occupied wheelchair are important to understand, for instance when judging if a door is wide enough to pass through, if there is enough space in which to turn around, or if there is enough clearance beneath the wheelchair to pass over an object on the ground.

Getting into, out of, and repositioning oneself with respect to the wheelchair: This includes transferring between the wheelchair and various other surfaces, unloading pressure-sensitive body parts, and changing position in the wheelchair.

Moving the wheelchair around on smooth level surfaces: Although the method of propulsion may vary, depending upon the impairments of the wheelchair user (e.g., using two hands, one hand and one foot, power), basic propulsion includes being able to move the wheelchair forward and backward, being able to turn in place or while moving, and being able to maneuver the wheelchair into position (e.g., to pick something up off the ground, getting close enough to a bed to make a transfer, negotiating doors).

Using the environment: Although the environment is often a barrier to activities, there are times when it can be an asset, especially for manual wheelchair users. For example, when turning a manual wheelchair around a solid object, placing a hand on the object can allow the wheelchair to swing around the object without slowing down, rather than the usual approach of slowing down and turning using both hand-rims. Other examples are when the wheelchair user uses the hand rails on a ramp to pull the wheelchair up the slope or uses a doorframe to guide passage through a door.

Skills that require leaning in the wheelchair: The wheelchair user's position in the wheelchair has a dramatic effect on the amount of weight that is on the front versus rear wheels because the wheelchair user's trunk and upper body constitute a considerable proportion of the combined mass of the wheelchair and wheelchair user. This is especially true for manual wheelchairs. Leaning to alter weight distribution with respect to the wheels will affect the stability of the wheelchair in a predictable way. For instance, when ascending an incline in a manual wheelchair, there is a risk of the wheelchair tipping over backward. To prevent this, the wheelchair user should lean forward enough to keep the front wheels on the surface.

In addition to stability, the balance of weight between the front and back wheels has an effect on rolling resistance. Wheels with large diameters have lower rolling resistance, whereas small-diameter wheels will tend to dig into soft surfaces. When crossing soft surfaces (e.g., carpet, gravel, grass), the wheelchair user should keep his/her weight on the rear wheels to the extent possible. When crossing side slopes, the tendency for the wheelchair to turn downhill can be reduced by leaning away from the swivel casters. If Leaning toward one side can also affect the lateral stability of the wheelchair. Also, if one wheel is spinning due to a lack of traction, this can often be corrected by leaning toward the spinning wheel.

Skills that require popping the front wheels briefly off the surface: There are some obstacles that require that the smaller (usually front) wheels clear the obstacle. These skills are most appropriate for manual wheelchairs. Examples include negotiating gravel, potholes, vertical obstacles (e.g., door thresholds), and getting up level changes (e.g., curbs).

Skills for which balancing on the rear wheels is necessary: For manual wheelchair users, the full wheelie position (balancing on the rear wheels) can be used to deal with situations like those described above, which require that the front wheels be unloaded. However, there are some desirable skills that can only be carried out by keeping the front wheels off the surface. These skills include the stationary wheelie (e.g., to improve neck comfort), turning around in a tight space, the forward descent of large level changes (e.g., a high curb), and the forward descent of steep inclines. These skills require the ability to perform a stationary wheelie, to turn around in the wheelie position, and to move forward or backward in the wheelie position. These skills are impossible in most powered wheelchairs and scooters.

Working with a helper: Most wheelchair users have at least some skills that they cannot safely perform themselves or that they find stressful. In such situations, the wheelchair user can benefit from the assistance of a helper. This may be in the form of minimal assistance (e.g., someone standing nearby to respond to a tip), the caregiver doing the task completely (e.g., ascending a curb), or the caregiver working in combination with the wheelchair user. The helper may be a regular one (e.g., friend, family member) or a passerby who can be recruited to help under the wheelchair user's direction.

1.17 WSP forms

The forms that facilitate the administration, recording, and reporting of each of the five versions of the WSP can be found at www.wheelchairskillsprogram.ca/eng/manual.php and in Appendices 3 through 5.

13

Chapter 2 Introduction to the assessment of wheelchair skills

There are a variety of measures that can be used to assess wheelchair skills, a comprehensive discussion of which is beyond the scope of this book. However, it may be helpful to consider the available measures as ranging along a spectrum of granularity from less- to more-detailed measures.

At the less-detailed end of the spectrum, there are questionnaire-based measures. One of these, the Wheelchair Skills Test Questionnaire Overview Scores (0–4 scores reflecting the subject's overall perception of his/her wheelchair skills capacity, confidence or performance) can be found on the WSP website (http://www.wheelchairskillsprogram.ca/eng/documents/WST-Q_OS.1.pdf). Another example of a low-granularity questionnaire is the Life Space Assessment Score that provides an ascending score corresponding to being limited to the room where one sleeps, being in other rooms of the home, being outside the home, being in the neighborhood, being outside the neighborhood, and being outside one's town.

More technology-based measures at the low-granularity level are data loggers (e.g., to document the number of kilometers travelled in a day or the number of times the tilt mechanism is used) and global positioning system sensors (e.g., to document where the wheelchair travelled during the day).

At the very detailed granularity level, examples of assessment tools are the use of instrumented rear wheels to document the forces applied to the hand-rims, the Wheelchair Propulsion Test (to assess such parameters as cadence, push efficiency) (see http://www.wheelchairskillsprogram.ca/eng/propulsion_test.php), video-recordings, three-dimensional motion analysis to document the relative movement of body parts, and oxygen consumption studies to document the metabolic energy cost of wheeling.

The WST and WST-Q that are the primary assessment methods covered in this book are measures that focus on the intermediate level of granularity. These measures test a subject's ability to perform a representative set of skills and, in the case of the WST-Q, current confidence in performing the skills and how often these skills are performed. Arguably, this intermediate level of detail is the level of greatest interest to wheelchair users, their caregivers, and their clinicians. Knowing such details provides the data needed for intervention through a change in wheelchair type, wheelchair set-up, skills training, modification of the physical environment, or provision of needed assistance. The WST and WST-Q are not intended to serve as "readiness" tests for independent wheelchair use, although they may be components of such an assessment.

Which measure(s) should be used to assess wheelchair skills depends upon the purpose of the assessment, the measurement properties (e.g., reliability, validity) of the tool, the characteristics of the test subject, the features of the wheelchair, the propulsion method, the equipment necessary, the skill of the assessor, and the time available. However, for the purpose of this book, the emphasis will be on the WST and WST-Q.

Chapter 3 The Wheelchair
Skills Test (WST)

The WST is a standardized evaluation method that permits a set of representative wheelchair skills to be simply and inexpensively documented. This test is intended to assess a specific person in a specific wheelchair in a standardized manner.

As noted earlier, the measurement properties of the WST have been studied to a moderate extent (see http://www.wheelchairskillsprogram.ca/eng/publications.php). In these studies, the WST has been found to be safe, practical, reliable, valid, and useful. The WST has been used as a screening or outcome measure in a number of studies. Further study is needed to re-evaluate the measurement properties of the WST as it evolves, in different settings and with different clinical populations.

3.1 Setting and equipment needed

The test setting for the WST should be reasonably quiet, private, free from distractions, and well lit. A standardized obstacle course may be used, but is not necessary. The specifications for such an obstacle course are provided later in this book in the chapter on individual skills. Some of the tests (e.g., "turns controller on and off") require no equipment and can be performed anywhere. In general, the settings described in the sections on individual skills should be considered as guidelines to enhance standardization, rather than as rigid constraints. If lines are used to mark limits (e.g., during moving turns), whether it is permissible for the wheelchair parts in contact with the floor (or the subject's feet) to touch the lines depends on whether the inner or outer borders of the lines reflect the dimensions specified. For instance, if a skill setting states that the subject must stay within a 1.5-m-wide space, if the outer borders of the lines represent the 1.5 m dimension then the subject may touch the line. Comparable challenges in the existing natural or built environment (e.g., in and around a hospital, the wheelchair user's home), may be used. Indeed, the WST can be completed as part of a community outing. However, if the setting is materially different from the one specified, this should be noted in the Comments section of the WST Form and may preclude the WST values from being compared to those conducted in more standardized settings.

3.2 Indications

For clinical purposes, the WST can be used early in the course of a rehabilitation program as a diagnostic measure, especially to determine which (if any) skills might be addressed during the rehabilitation process (e.g., by training, equipment change). However, predicting future performance on the basis of early attempts is of limited use. The trainer should not prejudge the outcome of training. By repeating the test on completion of the rehabilitation phase (or later during follow-up), the WST can be used as an outcome measure. The WST may also be used for program evaluation, to answer research questions, and to assist in wheelchair design.

3.3 Contraindications

No skill should be objectively evaluated if the subject is unwilling to attempt it or if the subject or WSP personnel would be placed at undue risk during testing (e.g., due to the subject's unstable cardiac disease, uncontrolled seizures, excessive weight).

3.4 General instructions to test subject

The paragraph below may be paraphrased or read to wheelchair-using subjects when the WST is being administered. It can be modified slightly if the subject is a caregiver or if the purpose of the WST is research.

"For about the next 30 minutes, I will be asking you to perform a number of different skills in your wheelchair. The reason for this is to find out which skills you do well and which might benefit from some practice or from changes to your wheelchair. We want to see if you can perform the skill properly and safely. We do not want you to hurt yourself, but there are some mild risks involved. To reduce the chances of you hurting yourself, we will be spotting you while you try each skill. Please wait until the spotter is in position before attempting each skill. The spotter will say "spotter on" to indicate when he/she is in position to protect you and "spotter off" to indicate if he/she is no longer in position. Please do not overexert yourself. We do not expect you to be able to perform every skill. Please do not try any skill that you are not comfortable performing. If you do not understand what we are asking you to do, feel free to ask questions. There is no need to hurry; this is not a race. If you would like to take a rest or to stop at any time, feel free to tell us. Do you have any general questions now, before we begin?"

Instructions may include gestures for people with language disorders or be in writing for people with hearing disorders but the tester should not demonstrate the skill. When giving instructions for each skill, before moving into the best position for observing and spotting the skill (if the tester is also serving as the spotter), the tester should stand or sit to the front or side of the subject so that the subject can see and hear the tester well. The tester must not instruct the subject in *how* to accomplish the task. If the tester asks for the task to be performed on both the left and right sides (e.g., turning the wheelchair around) but the subject performs the skill on only one side, the tester may prompt the subject (e.g., *"Now in the other direction"*) without penalty.

3.5 Feedback

After the attempt, nonspecific feedback may be given on how the subject did—for instance, *"You did well."* If the subject fails a skill, neither feedback on the reason for the failure nor instruction on how the skill might have been performed better should be given prior to completion of the entire WST. To do so would not affect the score for the skill already tested, but there may be other skills later in the WST that could be influenced by premature instruction. If observers (e.g., students, family members) are present during the test, they should be asked to remain silent and to refrain from providing cues or feedback. Once the entire WST has been completed, the tester may explain the reasons for any failures unless the WST is being administered to a research participant.

3.6 Disclaimer re sensitivity and specificity

The WST is a sensitive and specific test. A change in the subject (e.g., by a reduction of spasticity), the subject's equipment (e.g., removal of a prosthesis), a change in the wheelchair (e.g., by addition of a RAD), and/or a change in the test environment (e.g., by lowering lighting conditions) may affect the test scores. The objective WST findings are therefore specific to the situation assessed.

3.7 Number of attempts permitted

For each skill during the WST, the subject is ordinarily permitted only a single attempt. During the course of any single attempt, a subject may use different approaches (e.g., in a manual wheelchair first attempting the soft-surface skill forward, then backward if unable to proceed or in a powered wheelchair, pausing to change controller settings, the degree of tilt). It is only considered a second attempt if the subject clearly starts over (e.g., with a repetition of the instructions) and a significant pause between attempts.

There are some circumstances in which a second attempt may be permitted without penalty:

- If the subject misunderstood the instructions.
- If a correctable testing error (TE) is recognized when it occurs (e.g., the spotter intervened prematurely).
- If a subject appears to be rushing his/her skill attempts and failing to meet test criteria because of this, on the first occasion that this occurs, the tester may permit a second attempt and explain the importance of listening carefully to the instructions before beginning the skill attempt.
- It is sometimes the case that a test subject who has just failed a skill will ask for a chance to "try again." This may be permitted, but it is the first attempt that is scored.

A second attempt should not be considered a routine; ultimately, this is at the tester's discretion. If a second attempt is believed to be appropriate, the tester should provide no feedback on the reason for the failure on the first attempt, nor any instruction on how to perform the task between the two attempts. The task instructions may be repeated. If the skill is performed better on the second trial, the tester should record the better score. If a subject is unsuccessful when asked to perform a skill (e.g., "maneuvers sideways") but does it correctly later, incidental to another skill (e.g., the "level transfer"), the score must not be revised. The WST requires that the subject be able to perform the skill on demand. If a skill has been failed early in the attempt (e.g., after going through a door in one direction), it may still be useful to allow the subject to complete the remainder of the skill attempt (e.g., going through the door in the other direction) as a means of identifying issues that can be dealt with later during training.

3.8 Use of aids

Aids (e.g., for reaching) are permitted if the subject carries them with him/her. An animal (e.g., a service dog) that assists with the performance of a skill is considered an aid, not a caregiver, for the purpose of the WST.

3.9 Scoring of individual skills on capacity

The tester scores the success in accomplishing each skill, using the scale shown in Table 3.1. If there are criteria specific to individual skills, these are noted later, in the chapter on individual skills.

3.10 Comments

In addition to the capacity score for each skill, the comments add valuable qualitative data to the WST. The tester should record any comments that are appropriate (e.g., the reasons for any failures, left–right asymmetry). If there is appropriate spotter intervention during a skill attempt, the extent of the intervention and the reason for it should be recorded in the Comments section. The extent of spotter intervention may consist of a warning to a subject to stop or change the approach, minor physical contact from the spotter (even if the subject was able to complete the trial), or full intervention (e.g., if the subject required the spotter to prevent him/her from potentially injuring him/herself).

The nature of any potentially dangerous incident should be documented. Note should be made of any observations that require action (e.g., further training in alternative ways to accomplish a task, a change in equipment that might help). The WST tester should be alert to potentially correctable limiting factors in the wheelchair user's health (e.g., limited range of motion), wheelchair (e.g., rear axles too far back), and environment (e.g., if the WST is performed in the subject's home, a doorway that is too narrow). Comments by the test subject may also be recorded.

3.11 Training goals

If, at the beginning of the WST, it is decided by the tester or subject that one purpose of the assessment is to identify potential training goals then, before the assessment of individual skills, the subject should be asked whether there are any specific wheelchair skills on which he/she would be interested in receiving training. Doing this before assessing the individual skills is intended to reduce the likelihood of "training to the test." After the objective assessment of each skill has been completed (regardless of the scores recorded) and if an assessment of training goals is one of the purposes of the assessment, the subject may be asked whether that skill is one for which he/she would like to receive further training. On completion of the assessment of individual skills, the subject is asked whether there are any other skills on which he/she would be interested in receiving training.

3.12 Timing

The WST only requires the timing of 3 skills—"rolls forward short distance," "relieves weight from buttocks," and "performs stationary wheelie." These need only be timed to the nearest second. However, the time required to perform other individual skills, a series of skills, or the entire WST can provide an additional level of sensitivity to change (e.g., due to training or the use of a different wheelchair) that clinicians or researchers may wish to use.

Table 3.1 Scale for Scoring Skill Capacity Objectively on the WST

Pass (Score of 2):

- Task independently and safely accomplished without any difficulty. Unless otherwise specified, the skill may be performed in any manner. The focus is on meeting the task requirements, not the method used. Aids may be used.
- A "pass" score may be awarded automatically if the subject has passed a more difficult version of the same skill (e.g., if a subject successfully "ascends high curb," a pass may be awarded on the "ascends low curb" skill without the subject needing to actually perform the latter).

Pass with difficulty (Score of 1):

- If the evaluation criteria are met, but the subject experienced difficulty worthy of note (e.g., excessive time or effort required, inefficient method used, ergonomically unsound method used, poor technique that may lead to overuse injury at a later time, minor injury [e.g., minor blisters, abrasions] incurred), or a caregiver creates more than minimal discomfort or potential harm.
- Unintentional transient tips are not sufficient reasons to fail a subject's attempt at a skill but may justify a "pass with difficulty" score (e.g., If they interfere with smooth completion of the skill, are startling to the wheelchair user).
- If a tester is ambivalent about whether to award a score of 1 versus 2, he/she may find it helpful to ask him/herself if the identified problem is one that warrants efforts to resolve.

Fail (Score of 0):

- Task incomplete.
- If there are defined limitations of the space within which the skill is to be performed and the wheelchair wheels or the subject's feet in contact with the ground extend beyond those limits. Feet on footrests or wheelchair parts not in contact with the ground are usually permitted to extend beyond the limits, to simplify testing.
- Unsafe performance. A skill is considered unsafe if the subject requires appropriate and significant spotter intervention to prevent acute injury to the subject or others. A significant spotter intervention is one that interferes with the skill performance. Full tips should never occur, because the spotter should intervene. A skill performance is obviously unsafe if it results in a significant acute injury (e.g., lacerations, sprains, strains, fractures, head injury) that interferes with test continuation. Performing a skill quickly is not, in and of itself, unsafe. If the spotter intervention is one that neither hinders nor helps the subject, it can be ignored ("no harm, no foul").

(Continued)

Table 3.1 (*Continued*) Scale for Scoring Skill Capacity Objectively on the WST

- Likely to be unsafe in the opinion of the tester (e.g., on the basis of the subject's description of how a task will be attempted).
- Unwilling to try.
- Has failed an easier version of the same skill (e.g., if the subject fails the "ascends slight incline" skill, he/she automatically fails the "ascends steep incline" skill).
- If a caregiver is the subject of testing, he/she may not ask the wheelchair occupant for advice or physical assistance in the performance of the skill unless specifically permitted in the caregiver section of the individual skill descriptions or unless a blended WST is being administered.
- Wheelchair part malfunction preventing completion of the skill.

Not possible (score of NP):

- The wheelchair does not have the parts to allow this skill. For instance, if a manual wheelchair does not fold, the "folds and unfolds wheelchair" skill cannot be tested.

Testing error (score of TE):

- If the tester cannot assess the skill for some reason. For instance, the tester may not be able to get the wheelchair user into position to test the skill (e.g., on the floor for the "gets from ground into wheelchair" skill) or if a necessary item of equipment (e.g., the battery charger for a powered wheelchair is not available).
- If testing of the skill was not sufficiently well observed to provide a score (e.g., if the skill is being scored from videotape and the entire skill could not be viewed).
- If a correctable TE is recognized when it occurs, the test should be repeated.
- If there is a minor TE that the tester judges as not affecting his/her ability to score the test, this can be ignored.

There is no formal upper time limit for each skill or for the entire WST. This is to avoid the necessity of the tester timing every skill and to avoid having the subject feel rushed to complete the task. Although in real life a skill must be performed within a practical time to be useful, the definition of what such a time limit should be may vary with the circumstances. Fortunately, when administering the WST, this does not usually present a dilemma because the subject stops a task when it is taking too long. However, if a subject is persistently taking an apparently hopeless approach, the tester may intervene and stop the testing of that skill.

3.13 Rests and breaks

Rests are permitted during the skill attempts, unless precluded by the nature of the skill (e.g., the "performs stationary wheelie" skill). If the subject is making progress, he/she should be allowed to continue. Resting and then continuing is not considered a second attempt. For instance, a subject may get the casters up on the low curb, rest for a moment, then get the rear wheels up on the curb. It is also permissible for subjects to rest between skills. Indeed, there is no need for all of the skills to be performed on the same day. The WST is a test of individual skills, not a test of endurance. However, if the testing is conducted on more than one day, the tester should document the dates. Also, the wheelchair, its set-up, and subject aids (e.g., prosthesis) must remain the same on both test occasions if an overall score is to be valid.

3.14 Order of tests

During the WST, the tests may be performed in any order. For instance, it is usually practical to test the subject's ability to fold and unfold the wheelchair after testing the ability to transfer out of the wheelchair, but before evaluating the transfer back into the wheelchair. The order of testing may also vary depending on the availability and layout of equipment and test settings. For highly skilled test subjects, it may even be practical to use a "top-down" approach, starting with the more advanced of similar skills. As noted earlier, if the subject can perform the advanced-level version of some skills (e.g., the "ascends high curb" skill), then a pass may also be awarded for the simpler version of the same skill (e.g., the "ascends low curb" skill).

3.15 Left- versus right-sided components of skills

In objectively evaluating skill performance, both sides are tested (e.g., turning to left and right). Although this may be redundant for subjects with symmetrical impairments (e.g., of strength, range of motion), it may be valuable for subjects with asymmetrical impairments (e.g., due to hemiplegia, amputation) or for wheelchairs with asymmetrical flaws (e.g., a bent wheel rim on one side). A left-sided skill can be performed using the right hand without penalty and vice versa.

3.16 Minimizing ways in which training can invalidate WST scores

There are three avoidable ways in which wheelchair skills training can have undesirable effects on WST scores:

Inflation of the baseline score: If the same person is serving as both the tester and trainer, he/she may be tempted to conduct testing and training together. For instance, if the subject fails the "gets over threshold" skill, the tester/trainer may be tempted to provide instruction immediately, before continuing with the testing. However, the tester should complete as much of the pre-training WST as possible before beginning any training because the pre-training score of some skills may be artificially inflated by just having received training on a similar skill. In the threshold-skill example, training is likely to improve the subject's ability to perform the subsequent "gets over gap" skill. To reduce potential frustration by a subject who wants to proceed immediately to training, the tester should explain the process and indicate when training on the skills will be provided.

Failure to ensure skill retention: It is not unusual for a subject learning a new skill to experience transient success during a training session (skill "acquisition"), but to be unable to perform the same skill at the next session (skill "retention"). The ultimate goal of training is that the subject will be able to perform the skill in a variety of settings at any time in the future (skill "transfer"). To ensure at least short-term retention, the post-training WST should be performed at least 3 days after the training has been completed.

The "training to the test" or "specificity of training" phenomenon: If the training and testing are carried out in the same setting, it is possible that the subject may perform well in that setting, but not others. The trainer should be aware of this phenomenon, should have the subject practice in a variety of settings (e.g., indoors, outdoors), and should vary the order of skills during practice after initial acquisition and retention. This increases the likelihood that the subject will be able to transfer or generalize the skill.

3.17 Calculated scores

The following scores can be calculated by hand (as described below) or by using software developed for the purpose. Subtracting the number of NP scores from the denominator avoids penalizing test subjects by the inclusion of skills that would be impossible to complete. Subtracting the number of TE scores has a similar purpose. However, there may not be more than 2 TE scores for a calculated score to be valid.

Total WST Capacity Score (%): The formula is shown below. Possible percentage scores range from 0% to 100%.

Total WST Capacity Score = sum of individual skill scores/([number of possible skills − number of NP scores − number of TE scores] × 2) × 100%

Goal Attainment Score (%): Goal setting is discussed later in the section on training. The Goal Attainment Score (GAS) is of use when only a limited number of skills are addressed, through training. The numerator is the number of skills that are met and the denominator is the number of goals set. The formula below is based on a simple yes/no score for each skill, but alternative scoring (e.g., yes/partially/no) can be used. The GAS at baseline is 0% by definition. Possible percentage scores after intervention range from 0% to 100%.

Goal Attainment Score = number of goals met/number of goals set × 100%

Special Purpose Scores (optional): Any subset of individual skills may be selected for a calculated subtotal percentage score. For instance, the scores for individual skill levels (i.e., basic, intermediate, advanced) can be calculated. Other examples are a score that deals only with skills that might be appropriate for foot-propellers of manual wheelchairs or a blended score that reflects the combined efforts of a wheelchair user and a caregiver. If the WST is being used with a highly skilled person or group, there may be a ceiling effect (scores near 100%). In such a situation, the WST can be extended in ways that allow change to be detected. For instance, a skill or group of skills can be timed. Alternatively, skills can be carried out in combination (e.g., wheelie turn in place on a soft surface). The criteria can be made more difficult (e.g., turning in place in a 1.0-m-square space instead of a 1.5-m-square one. Any such modifications should be documented so that the results can be interpreted.

3.18 WST test report

There is one WST form (Appendix 3) for each of the five versions of the WST. The WST form may be completed by hand or be generated by software. The completed WST form includes identifying data, the scores for individual skills, the calculated capacity score, comments, and the skills (if any) for which the subject would be interested in receiving training.

Chapter 4 The Wheelchair Skills Test Questionnaire (WST-Q)

The relationship between the WST and the WST-Q has been reported in the literature (see http://www.wheelchairskillsprogram.ca/eng/publications.php). The correlation between the total WST and WST-Q scores has been found to be high, although the WST-Q scores are slightly higher. The WST and WST-Q each have advantages and limitations as summarized in Table 4.1.

The advantages of the WST-Q include that it requires less time, equipment, and space to perform, it does not appear to induce much of a training effect (like the WST seems to do), avoids a training-to-the-test effect, allows one to assess confidence and performance as well as capacity (in the terms of the International Classification of Functioning, Disability, and Health), is more realistic (relating as it does to the subject's own setting), is not subject to limitations due to missing equipment (e.g., a battery charger) at the time of testing, subjects are not likely to fail a skill on a technicality (e.g., a wheel slightly over a line), the settings are less specifically defined, and the WST-Q may be the only option for situations in which objective testing is impractical or impossible (e.g., during telephone interviews). The WST-Q can be administered by phone, postal questionnaire, or online. It can be completed by a proxy. There is no risk of injury.

The limitations of the WST-Q are that the tester must rely on the subject's ability to understand the questions and to communicate valid answers. This limitation can be offset by having a proxy (e.g., a caregiver) who knows the subject well or a translator assist in providing the answers. There is potential for the subject to overestimate or underestimate his/her capacity and performance. The WST-Q does not provide any detail about *how* the skills are performed, limiting its usefulness as a guide to intervention (e.g., by altering the wheelchair set-up, by training).

The complementary benefits of the WST and WST-Q can be captured by using them in combination—"*Can you do it? How confidently? How often? Show me how.*"

Table 4.1 Comparison of WST and WST-Q

Consideration	WST	WST-Q
Time to administer	~30 minutes	~10 minutes
Obstacles needed	Yes	No
Space needed	~1000 square feet	None
Induces a training effect	Probable (~5%)	None known
Can assess capacity (*can* do)	Yes	Yes
Can assess confidence	No	Yes
Can assess performance (*does* do)	No	Yes
Simulated versus real setting	Simulated usually	Real
Affected by missing equipment	Yes	No
Likelihood of failing a skill on a technicality	Occasional	None
Degree of specificity of settings	High	Low
Possibility of a TE	Occasional	Rare
Can be administered by phone	No	Yes
Can be administered by mailed questionnaire	No	Yes
Can be administered online	No	Yes
Can be completed by a proxy	No	Yes
Requires ability to follow instructions	Yes	No
Requires ability to communicate	No	Yes (unless proxy)
Potential to misrepresent functional level	Low	Slightly greater
Provides detail about *how* the skills are performed	Yes	No
Risk of injury	Minimal	None
Total scores	Slightly lower	Slightly higher

4.1 Indications

As for the WST, for clinical purposes, the WST-Q can be used early in the course of a rehabilitation program as a diagnostic measure, especially to determine which (if any) skills might be addressed during the rehabilitation process. By repeating the test on completion of the rehabilitation phase (or later during follow-up), the WST-Q can be used as an outcome measure. The WST-Q may also be used for program evaluation, to answer research questions and to assist in wheelchair design.

4.2 Contraindications

The WST-Q is only valid if the subject (or proxy) is able to communicate. As a screening procedure, the tester should ask the potential subject about information (e.g., date of birth, diagnosis, length of time using a wheelchair, time up in the wheelchair each day) that can be confirmed by chart review, the nursing staff, or family members.

4.3 Time limits

There is no upper time limit for the WST-Q. Rests are permitted but are usually unnecessary because the average time to complete the WST-Q is only about 10 minutes. If the testing is conducted on more than 1 day, the tester should document the dates.

4.4 General template for WST-Q individual skill questions

For individual skills, the initial question is about capacity. The capacity question, answer options, and definitions are summarized in Table 4.2. A score for this question is mandatory for each skill.

The next question about each individual skill is about confidence. Confidence in one's ability is an important determinant of the extent to which wheelchair skills are actually used in everyday life. The answer options and definitions are summarized in Table 4.3. The confidence questions are optional and may be skipped if an assessment of confidence is not one of the purposes of the questionnaire.

If the answer to the capacity question for a skill is "no," the score on the confidence question is automatically 0. If the score for capacity is NP, then NP is automatically the score for confidence. If there has been a TE for the capacity question, the confidence score is TE by definition.

The next question about each individual skill is about performance. The answer options and definitions are summarized in Table 4.4. The performance questions

Table 4.2 WST-Q Capacity Question, Answer Options, and Definitions for Each Skill

Capacity Question: "*Can You Do It?*"		
Answer	**Score**	**What This Means**
Yes	2	I can safely do the skill, by myself, without any difficulty.
Yes with difficulty	1	Yes, but not as well as I would like.
No	0	I have never done the skill or I do not feel that I could do it right now.
Not possible with this wheelchair	NP	My wheelchair does not have the parts to allow this skill. (This option is only presented for skills where such a score is a possibility.)
Testing error	TE	When answers have not been recorded (e.g., inadvertently or because the test subject did not understand the question).

Table 4.3 WST-Q Confidence Question, Answer Options, and Definitions for Each Skill

Confidence Question: "*How Confident Are You?*"		
Answer	Score	What This Means
Fully	2	As of now, I am fully confident that I can do this skill safely and consistently.
Somewhat	1	As of now, I am somewhat confident that I can do this skill safely and consistently.
Not at all	0	As of now, I am not at all confident that I can do this skill safely and consistently.

Table 4.4 WST-Q Performance Question, Answer Options, and Definitions for Each Skill

Performance Question: "How Often Do You Do It?"		
Answer	Score	What This Means
Always	2	Whenever I need or want to do so.
Sometimes	1	Sometimes when I need or want to, sometimes not.
Never	0	Never or less often than once a year.

are optional and may be skipped if an assessment of performance is not one of the purposes of the questionnaire.

WST-Q performance is related to WST-Q capacity, but is also related to personal factors (e.g., age, confidence) and the environment (e.g., weather, architectural barriers, opportunity). Additionally, some skills (e.g., "folds and unfolds wheelchair," "gets from ground into wheelchair") may not need to be performed frequently. Total capacity percentage scores tend to exceed total performance percentage scores. The converse could occur—for instance, if a subject had a number of skills for which the capacity scores were "pass with difficulty" but these skills were always performed when necessary. Also, if a wheelchair user had an acute injury (e.g., a fractured wrist), he/she might be unable to perform a skill currently that he/she had always performed in the past. However, there is no guarantee in such a circumstance that the wheelchair user will ever get back to the earlier level of performance. Therefore, for the purposes of the WST-Q, if the capacity score for an individual skill is 0, the performance score for that skill is also automatically 0. If the score for capacity is NP, then NP is automatically the score for performance. If there has been a TE score for the capacity question, the performance score is automatically TE.

If the purpose of the WST-Q includes being able to provide an overall score for each skill and in total, a composite score can be calculated. The composite score

Table 4.5 WST-Q Training Goal Question, Answer Options, and Definitions for Each Skill

Question: "*Is This a Training Goal?*"	
Possible Answers	**What This Means**
Yes	I am interested in receiving training for this skill
No	I am not interested in receiving training for this skill

is the sum of the capacity, confidence, and performance scores (i.e., a range of 0–6 for individual skills and 0%–100% for total percentage scores).

As for the WST, at the beginning of the WST-Q, if it is decided by the tester or subject that one purpose of the WST-Q is to identify potential training goals then, before the assessment of individual skills, the subject is asked whether there are any specific wheelchair skills on which he/she would be interested in receiving training. After the capacity, confidence, and performance questions have been answered (regardless of the scores recorded) and if an assessment of training goals is one of the purposes of the assessment, the final question for each skill is about training goals. The goal question, answer options, and definitions are summarized in Table 4.5.

On completion of the assessment of individual skills, the subject is asked whether there are any other skills on which he/she would be interested in receiving training.

4.5 Scoring algorithm for individual skill questions

The algorithm for the individual skill questions is shown in Figure 4.1.

4.6 Calculated scores

The total scores for capacity that can be calculated from the WST-Q data are identical to those for the WST described earlier. Additionally, for the WST-Q total confidence, performance, and composite scores can also be calculated as follows:

Total confidence score = sum of individual skill scores/([number of possible skills – number of NP scores – number of TE scores] × 2) × 100%.

Possible percentage scores range from 0% to 100%.

Total performance score = sum of individual skill scores/([number of possible skills – number of NP scores – number of TE scores] × 2) × 100%

Possible percentage scores range from 0% to 100%.

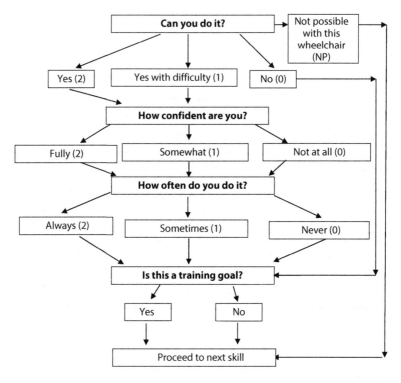

Figure 4.1 Scoring algorithm for individual skill questions.

Total composite score = sum of individual skill composite scores/ ([number of possible skills – number of NP scores – number of TE scores] × 6) × 100%.

Possible percentage scores range from 0% to 100%.

4.7 Options for how the WST-Q may be administered

There are a variety of acceptable ways in which the WST-Q can be administered and recorded. The WST-Q may be tester administered in-person or by telephone with the tester reading the questions and recording the answers (Appendix 4). If a tester is involved, he/she may explain a question if it is not understood by the subject. The tester may also use follow-up questions to reassure him/herself about the validity of the answers provided (i.e., a semi-structured interview).

Alternatively, the WST-Q may be self-administered (e.g., in a postal or online questionnaire) with the test subject or proxy reading the questions and recording the answers. For the paper version of the WST-Q, the person administering the questionnaire may record the answers either on the WST-Q script or on the WST-Q report form.

A computer-assisted version of the WST-Q for desktop computers and tablets is available, links to which are provided on the website. The tester or test subject records the answers on the computer or tablet. The advantages of this approach are that instances of missing data and transcription errors are minimized. Also, it requires less time to complete the WST-Q in this way because the computer uses the scoring algorithm automatically.

The WST-Q and WST can be administered together. For each individual skill, the questions about capacity, confidence, performance, and goals can be followed by a demonstration of that skill. Alternatively, the WST can be administered after the full WST-Q is completed.

4.8 WST-Q report form

There is one WST-Q report form (Appendix 5) for each of the five versions of the WST-Q. The WST-Q report form may be completed by hand or be generated by software. The completed WST-Q report form includes identifying data, the scores for individual skills, the calculated score(s), comments, and the skills (if any) for which the subject would be interested in receiving training.

Chapter 5 Wheelchair skills training

The WSTP represents the WSP developers' attempt to combine the best available evidence on motor-skills learning ("process") with the best available evidence on how to perform specific skills ("content").

The optimum way to perform and teach each wheelchair skill may vary depending upon the characteristics of the learner, the wheelchair being used, and the setting. However, the WSTP-training protocol uses training methodology based on the literature. Research evidence regarding the safety and efficacy of WSTP training is available in Chapter 9.1 and at http://www.wheelchairskillsprogram. ca/eng/publications.php. Although much further study is necessary, in all of the studies to date, WSTP training has been found to be safe, practical, and to result in significantly greater improvements in wheelchair skills than standard care. There is also research evidence in the literature for some of the specific skills (e.g., basic propulsion technique, transfers, inclines, curbs, wheelies) but not for all skills.

5.1 General background on motor skills learning

Education can address one or more of three domains—knowledge, skills, and attitudes. All three are relevant to wheelchair skills training. However, in this section, the focus will be primarily on motor skills learning.

The issues presented in this section are based on the extensive motor-skills-learning literature (over 500 English language articles published each year) and on the experience of the WSP team. This chapter is not intended to be a treatise for researchers. It is an attempt to synthesize the aspects of this literature that are most relevant to the learning of wheelchair skills. The attempt has been to express these principles in language that the average educated (but not necessarily professional) trainer and learner might understand. Although there is a great deal of scientific evidence underlying these principles, the principles themselves are fairly simple. Trainers and/or learners who understand and apply the principles will be more effective than those who do not. In addition to the general principles summarized in this chapter, more specific "training tips" are included with the individual skills later (see Chapter 7).

5.2 What is a "motor skill"?

A motor skill is one that is voluntary, observable, has been learned, and has a goal. Motor skills have been classified on the basis of the size of the muscle groups involved (gross vs. fine), on the basis of whether they are discrete tasks or more continuous ones, and on the basis of how stable the environment is (open vs. closed).

5.3 The learning process

In the course of learning a new motor skill, the learner progresses through various stages. This is sometimes referred to as the "learning continuum." Early in the process, success may be partial, inconsistent, or only possible in a familiar setting. As learning progresses, preliminary success is eventually achieved (skill acquisition), consistency within training sessions improves, success carries over into subsequent sessions (skill retention), and the learner is able to use the skill in more diverse settings (skill transfer). Ultimately, the skill may become autonomous, requiring little or no conscious effort. The time course of motor learning includes an initial period of rapid improvement, sometimes followed by a plateau that may be followed by additional gains. The shape of the motor-learning curve is not linear and may be punctuated by abrupt transitions from novice to skilled coordination patterns.

There is a distinction between aspects of the learning process that are in the form of facts and ideas (sometimes called the "declarative," "cognitive," or "explicit" system) versus those that relate to the actual performance of the skill (sometimes called the "procedural," "motor skill," or "implicit" system). Each can be acquired without the other. If both are acquired, this need not be in a fixed order. The two can assist or interfere with each other. Attempting to consciously control motor actions can disrupt optimal performance. Skills learned implicitly through a discovery approach appear to be more robust under pressure. Healthy learners can sometimes engage explicit (conscious) and implicit (automatic) motor control simultaneously without deterioration of control compared to either alone.

People who have acquired expertise in performing a motor skill have some characteristics in common. For instance, they have greater awareness of their situations and better ability to anticipate changes in the environment. They are better able to exclude intrusions on their attention and to remain focused on the task. Their motor performances are less affected by stress and fatigue.

5.4 Assessment of wheelchair skills

Periods of formal evaluation (e.g., using the WST and/or WST-Q before and after training, and at follow-up) can be useful. In addition to the assessment measures mentioned earlier in Chapter 2, there are a variety of parameters that provide evidence of learning due to practice. Examples of such parameters are increased speed, improved consistency, improved adaptability to other settings, improved economy of movement, and improved ability to detect and self-correct errors. Ongoing assessment by the trainer is important. What the trainer can do to facilitate the learning process varies continuously. A training log may be used by the trainer and/or learner to track the training process.

5.5 Goal setting

From the baseline WST or WST-Q assessment, skills may be identified that are not performed as safely, effectively, or efficiently as they might be. Generally, only 5–10 goals should be identified at the beginning of a series of training sessions. The goal may be from the WSP skill set—a full skill, a part of a skill, a variation of a skill—or any other skill that is important to the learner. Goal pursuit is related to the learner's beliefs about him/herself and the task (confidence or self-efficacy). The learner may need help in coming to a decision about the goals of training because he/she may not initially recognize the functional benefits of acquiring a new skill. Additionally, a decision needs to be made as to whether it is feasible for the person to learn this skill. This is a judgment call and requires a good understanding of the learner's health and circumstances. If in doubt, it is recommended that the learner be given an opportunity to learn the skill. If progress is not being made, a learner can decide to abandon that skill. The trainer can assist the learner in coming to this decision.

Goals should be brief, specific, significant, achievable in the training time available, and observable (a measureable action item). A broad participation-level goal (e.g., to go shopping) can be broken down into the constituent skills that comprise it.

The following are examples of good WSTP goals:

- Roll 100 m in 2 minutes, using no more than 100 pushes.
- Get the wheelchair up a 2 cm curb.
- In the wheelie position, roll forward 10 m.
- Get from the wheelchair to the floor and back within 60 seconds.
- Come down a flight of 10 stairs backward, using one handrail.

The following are examples of poor goals from the perspective of WSTP training (and why):

- Go shopping at the mall (too broad, needs to be more specific).
- Reduce by 10% the number of pushes needed to roll 10 m (not significant).
- Complete a full marathon (may not be achievable in the training time available).

- Spend more time with my friends (not a wheelchair skill and not easily observable).
- Understand the importance of preventing pressure sores (not an action item).

Involving the learner in the goal-setting process can have a positive effect on motivation. However, the trainer has the right to refuse to provide training on any skill that he/she does not believe to be safe and feasible. The goals should be monitored and may be revised as training progresses. The goals may be formalized to allow a GAS (see earlier) to be calculated that can be used to track progress and quantify outcomes. A poster on Setting Goals is available at http://www.wheelchairskillsprogram.ca/eng/posters.php. It is intended to be printed and posted in the training area.

5.6 Individualizing the training process

Motor-learning principles generally apply almost equally well to elite athletes and to those who have severe disabilities. However, there is benefit to tailoring the training process to the learner. Learning-style preferences exist and should be respected whenever possible. Training can sometimes take the form of a problem-solving exercise, attempting to answer the question "For this learner, with this wheelchair, in this context, what would be the safest and most effective way to perform this task?" For another wheelchair user or another wheelchair, a different solution may be appropriate.

Inability to perform a skill may be due to a variety of limiting factors, alone or in combination. Limiting factors may be intrinsic (e.g., impairments such as cognitive limitations, weakness, pain, shortness of breath, limited range of motion, spasticity, poor coordination, movement disorders) or extrinsic (e.g., a faulty wheelchair part, poor seating support, poor lighting). The trainer should attempt to identify remediable limiting factors and seek to have them addressed.

Motor-skills learning can be affected by personal characteristics. A trainer who understands these differences will be able to reassure learners who might be progressing more slowly than others. For instance, males learn some skills faster than females. Although learning capacity is greater early in life and the young learn motor skills more rapidly and with less practice, aged people can acquire new motor skills well. Very young children learn better by practicing parts of skills but whole-skill practice works better by about the age of 10. Children acquire skills faster, perform them better and are more engaged when using scaled equipment. Motor learning may be affected by emotion or fatigue.

Neurological conditions may affect motor-skills learning. The impairments (e.g., motor weakness, spasticity, sensory loss, coordination difficulties, balance difficulties, perceptual problems) may affect how a skill should be optimally performed and the ease with which learning can occur.

Specific neurological disorders may also need to be taken into consideration, for instance:

- For people with stroke, the post-stroke brain has heightened sensitivity to rehabilitation early but this phenomenon declines somewhat with time. The extent of improvement is related to the intensity of training, but high doses of training may not be well tolerated early after the stroke. Explicit information disrupts skill acquisition even more than usual in people who have had strokes affecting the basal ganglia. For people with language impairments, it may be helpful to use nonverbal cues and feedback rather than verbal ones.
- People with multiple sclerosis may have greater susceptibility to high environmental temperatures and fatigue more easily.
- People with Alzheimer's disease can learn and retain new motor skills. Implicit-learning strategies and demonstration appear to be particularly useful for such people. Consistent practice conditions may work better than variable ones.
- For people with dementia, there is some evidence of superior learning of problem-solving tasks with the help of cues (errorless learning) versus trial-and-error learning.
- People with Parkinson's disease can learn new motor skills. Rhythmic auditory cues can be helpful for them. Although less helpful for healthy people, paying conscious attention to motor tasks can be useful for people with Parkinson's disease. Consistent practice conditions may work better than variable ones.
- Medicated patients with schizophrenia may have difficulties with the consolidation of skills.
- For children with cerebral palsy, a 100% feedback schedule is more effective than an intermittent one.

5.7 Structure of training

There are a variety of ways in which the safety, effectiveness, or efficiency of training can be enhanced. The motor-learning principles discussed in this chapter can be thought of as the trainer's "instructional tool kit" with specific tools to be used as needed. Training can take place anywhere (e.g., in the hospital, community, the learner's environment). Training can take place in an ad hoc format, seizing teaching opportunities as they present themselves (e.g., during community outings); although this approach has much to commend it after the individual skills have been learned, it is unlikely that such challenges will present themselves in the order that would be most helpful to optimize learning. In the clinical setting, it can be helpful to provide more structure (e.g., scheduled sessions with lesson plans). At the beginning of a training session, a warm-up can have a number of benefits. For instance, moderate intensity aerobic exercise improves motor skill acquisition. Sample lesson-plan templates for initial and subsequent sessions can be found in Appendix 1.

5.8 Training in pairs or groups

To permit an individualized approach, a ratio of trainers to learners of 1:1 or 1:2 is ideal, although ratios as low as 1:20 can be successful. Training in pairs or groups is practical, cost-effective, and has educational merit. The optimum group size depends on the goals but more than eight people in a group can lead to fewer interactions and lower satisfaction. Group training can permit group discussions and problem-solving. Learners can serve as models for each other, both for how and how not to perform a skill. Whenever possible, it is desirable to select groups on the basis of roughly similar skill level. Learners in groups should be reminded that skill capability is affected by a number of factors (e.g., age, sex, impairments, wheelchair type), so they should not compare their progress with that of others. For individuals with low self-efficacy, collaborative training with a more experienced partner aids skill acquisition. To function well, groups may need to reach an explicit agreement on the group process (e.g., avoiding the use of cell phones during the session) and consequences (e.g., singing a song) for breaking the group rules.

5.9 Motivation

Motor-skills learning is enhanced if the learner is motivated to learn. The trainer can help to motivate the learner by making the learning meaningful and rewarding. Game-based exercises can help to create and maintain interest. Children especially may learn best through play, rather than through formal training on a skill-by-skill basis. Working in either cooperation or competition with other learners can enhance motivation.

Whenever possible, the trainer should explain how the learner will benefit by learning a new skill. Training should be relevant to the learner and his/her context. In addition to the long-term benefits of training, there may be short-term benefits, such as the social interaction during the training sessions, the pleasure that some people get from challenging themselves or improving on a test. Without creating anxiety, the trainer should let the learner know that he/she will be assessed at the end of the training period, because this is known to have a positive effect on skill acquisition. Encouragement and positive feedback from the trainer or fellow learners can be powerful incentives as well. Rewards significantly enhance long-term retention of motor learning. The trainer should not be reluctant to challenge the learner to try ever more difficult but potentially achievable skills.

Learning, self-efficacy, and affect are better when the learner perceives him/herself as having a choice (e.g., *"Do you want to start at that end of the line of pylons or at this end?," "Would you prefer to wheel across the yellow or the green mat?"*). Autonomy is also important in deciding when and for how long to practice.

5.10 Demonstration

Demonstration is one of the most powerful motor-skills-learning tools. The demonstrator may be the trainer, a model, or a peer. Demonstration may be in-person or on a video. The Pictures and Videos section of the WSP website (http://www. wheelchairskillsprogram.ca/eng/specific_skills.php) contains numerous video clips that can be used. The demonstrator should ideally be skilled, but this is not a necessity. One approach is to use an expert model to provide an accurate template of the movement, followed by less successful models. If the model is at a similar level to the learner (e.g., in a group setting), the learner can learn from the feedback provided to the model.

If the learner is unfamiliar with the skill to be practiced, the demonstration should occur before practice begins. Otherwise, the demonstration can be used as part of the feedback provided to the learner. The demonstration may be repeated as often as needed. The trainer should briefly describe important elements of the skill or provide attention-directing cues, as part of the demonstration. The trainer should focus on what to do rather than what not to do, at least until the learner has had an opportunity to try the skill several times.

Observation alone can result in learning but has limits if not followed by physical practice. Demonstration is most effective for a novel task and less effective when refining a skill. When demonstrating a skill, the trainer should put equal emphasis on the movement and the outcome.

5.11 Verbal instructions

Using the terminology of the motor-learning literature, "instructions" are generally provided before practice, as distinct from "feedback" that is provided afterwards. Providing explicit instructions before task practice can be detrimental so instructions should be used with caution. Learners have a limited capacity to attend—the trainer should not overwhelm the learner with the quantity of information. Instructions are more likely to be of help for advanced learners (e.g., instructions regarding anticipation and decision-making). The length of time

41

between the instructions and actual practice should be minimized. Preferably, instructions should be given in combination with a demonstration. Learning is enhanced by instructions that portray the task as a learnable skill versus one that is based on inherent ability.

As for the content of instructions, some general examples are as follows:

- Speed and accuracy are inversely related. If both are desirable, the learner will do better to start with accuracy and build speed later. An instruction may be to *"Take your time, it's not a race."*
- The trainer may provide instructions about what to look for in the environment that might affect performance. For instance, *"Pay attention to the lip at the bottom of the ramp."*
- The trainer may provide a framework, an organization, or a way of thinking about a skill. An instruction may be *"Think of the right rear wheel of your wheelchair as the face of a clock and start with your hands at 11:00 o'clock."*
- Analogy learning has been found to be helpful. For instance, during the rolling forward skill, an instruction may be *"Coast between pushes just as you would between strokes when paddling a canoe."*
- The trainer may provide verbal cues—short, precise words, or phrases that direct attention or prompt movements. For instance, when attempting to get a manual wheelchair over a threshold from a stationary start, the trainer may ask the learner to *"Pop"* (popping the casters over the threshold) then *"Lean"* (leaning forward to help get the rear wheels over). The trainer should limit the number of cues to those that are most critical. It can be helpful to have the learner verbalize the cues prior to attempting the skill and during the attempt. As noted earlier, for people with dementia, there is some evidence of superior learning of problem-solving tasks with the help of cues versus trial-and-error learning.

5.12 Focus of attention

Early in training, the trainer may need to have the learner focus on specific actions or processes (e.g., *"Lean forward"*), if a crucial error has been identified. However, the research literature suggests that, when most individuals engaged in motor-learning tasks concentrate on movements themselves, the conscious intervention in the control processes results in poor performance and learning. People with Parkinsonism may be an exception to this general rule.

As the skill becomes more automatic, more advanced learners tend to do better if they focus on the overall goal or outcome of the skill performance (e.g., *"Get up the incline onto the platform*"). This phenomenon is better documented in adults than for children. Although automatic performance is ideal, even experts may find it necessary from time to time to focus attention on an aspect of a skill that requires it.

5.13 Imagery

There is evidence that imagery or mental practice can be helpful in the acquisition of motor skills. Imagery can be assigned as homework. Imagery can focus on what the learner would see during the performance of a skill, with internal or external perspectives (i.e., seeing through one's own eyes vs. seeing oneself as though watching another person). Alternatively, imagery can focus on what the person might feel (e.g., limb position, external forces) during a skill performance. Most studies have used verbal live or recorded imagery instructions, have been performed with the eyes closed and have used an internal perspective with a kinesthetic focus. On average, participants in such studies practiced for about 15 minutes at a time, 3 times a week for a total of about 3 hours. Even a short nap after motor imagery helps.

Imagery can also be used for motivational purposes (e.g., visualizing performance with confidence and ease). Imagery can be used in advance, to prepare to perform a skill, or after the attempt, to reinforce a well-performed trial. Imagery is not as effective as physical practice but it is better than no practice. Used in combination with physical practice, imagery is almost as effective as physical practice alone, so it may be a useful strategy when there are factors that prevent physical practice (e.g., bad weather, lack of spotter availability, a sore shoulder). Imagery has a greater effect on closed skills (ones that are always the same) than open ones. Imagery is less useful for a novel task than a familiar one.

5.14 Feedback

5.14.1 Types of feedback

Implicit learning through intrinsic feedback (e.g., from what the learner can see, hear, feel) is useful and may be all that is needed. Feedback can be augmented in a variety of ways (e.g., by watching oneself in a mirror [Figure 5.1], by watching a video of one's performance, by receiving biofeedback, by receiving feedback from a trainer). Augmented feedback is generally an effective tool for enhancing learning (e.g., by better participation, faster skill acquisition). However, augmented feedback is not always needed and it can hinder learning if the learner becomes dependent on it. The ultimate goal of skill learning is for the performer to be able to perform the skill without augmented feedback.

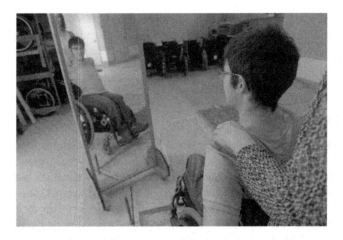

Figure 5.1 Example of augmented feedback, using a mirror to view caster position relative to a gap in the floor.

5.14.2 Feedback content

The trainer should be supportive and encouraging, even to the extent of slightly exaggerating how well the learner is doing in comparison with others at a similar stage of training. Such "bogus" positive feedback can have positive effects on skill acquisition, self-efficacy, and affect. However, the trainer should be accurate with respect to feedback content. It is counterproductive to tell a learner that his/her performance was successful if it was not. Most people learn at least as well from their failures as from their successes.

When learning wheelchair skills, feedback from the trainer about the success or failure of an attempt at a skill ("knowledge of results") is usually unnecessary, for two reasons. First, the result is usually self-evident. Second, if the learner is failing repeatedly, he/she may get discouraged by repeated statements about failure.

Another form of feedback is the provision of information about *how* the skill was performed ("knowledge of performance"). Ideally, such feedback should be directed at what the trainer suggests the learner should try differently ("prescriptive knowledge of performance"), in order to achieve a safer or more effective result. Before providing prescriptive knowledge of performance, it can be useful to ask the learner about his/her perceptions about the problem and intended solutions. The objective is to develop a learner who is an independent problem-solver. If the learner does not self-diagnose the problem correctly, the trainer should identify the most critical error and suggest what might be done to correct this problem.

Pointing out errors is more effective than noting what the learner is doing correctly (although the latter is important for motivation). It can be useful to have learners attempt skills in inappropriate ways (e.g., rolling across a soft surface while leaning forward, causing the casters to sink into the surface), to help them better understand why a suggestion is being made. Qualitative feedback is fine early in training (e.g., "*You need to pop your casters higher*"). Later, quantitative feedback (e.g., "*You need to pop your casters about 2 cm higher*") may be better. Feedback can be more effective if it directs the performer's attention away from his or her own movements and to the effects of those movements. The perceived expertise of the trainer (e.g., as evidenced by a demonstration of the skill being learned) affects the perceived usefulness of the feedback provided.

5.14.3 Timing of feedback

The optimum frequency for knowledge-of-results feedback (if any is needed) is affected by the difficulties of the task—the more difficult the task, the higher the frequency of feedback can be without interfering with skill acquisition.

When providing knowledge-of-performance feedback, the trainer needs to exercise judgment and to be attuned to the chemistry of the training session. The trainer should offer feedback statements no more often than after every second attempt. Autonomy can be provided ("*Let me know when you would like some feedback*"). An exception to this is if a learner performs in an unsafe manner and does not appear to be aware of it; the trainer should point this out as soon as possible. The trainer should let the learner know that the absence of feedback means that the performance was adequate for the current stage of learning. This gives the learner an opportunity to problem-solve on his/her own. It also decreases repetitive feedback statements, especially in the case of more advanced skills when it can take time for the learner to overcome a problem. A common error is for the trainer to spend too much time talking and not enough time allowing the learner to practice.

The feedback schedule is especially important for wheelchair users who have cognitive or behavioral impairments. A self-controlled feedback schedule (i.e., letting the learner ask for feedback) is generally preferable. The trainer should gradually reduce the frequency of feedback statements as time goes on. The feedback weaning schedule may need to be more gradual for children. As the fading process leads to less frequent feedback, the trainer should summarize a series of attempts rather than focusing only on the most recent attempt. This technique can also be used when working with a group, providing feedback that deals with a problem that several of the group members are encountering.

Trainers should be aware of the principles of behavior modification, which have similarities to the principles of motor learning. Positive reinforcement (e.g., an encouraging remark) increases the likelihood of a behavior (or skill) being performed, whereas negative reinforcement (or no reinforcement) has the opposite effect. Initially, the trainer's tolerance for the learner's errors should be broad, but the "bandwidth" of acceptable performance is gradually narrowed as learning proceeds. Behaviorists refer to this as "shaping" a behavior. Intermittent positive reinforcement, at irregular intervals, is the ideal reinforcement schedule for sustaining behaviors.

Feedback can be provided during the skill attempt. This is more practical for continuous skills (e.g., "rolls longer distance"), but there is a danger that this may interfere with the learner's attention to intrinsic feedback. Providing the feedback after the skill is usually preferable. The trainer should wait a few seconds before providing feedback to allow intrinsic processes to work first. Before beginning the next trial, the trainer should allow the learner some time to plan the next attempt. Any augmented feedback should be followed by an opportunity to practice.

Improvements in communication technology has made it possible for the learner and trainer to interact when separated in space ("remotely") and time ("asynchronously"). For instance, a learner in one part of the world who is having difficulty with a skill can send a video-recording of his/her technique to a trainer in another part of the world who can provide feedback at a later time that is convenient for the trainer. That feedback can be considered later, at a time convenient to the learner. The learner is not limited to an interaction with a single trainer but can seek input from anyone willing to provide it.

5.15 Specificity of practice

If a learner wants to improve his/her ability to perform a task, the task itself should be practiced. Cross-training may help to develop fitness, but is of limited use for the development of motor skills. However, there is mounting evidence, for a broad range of motor skills, that training in simulated situations can enhance skill performance in real-life situations. Practice should be as specific as possible with respect to the task itself and the context in which it is to be performed. If the goal is for the learner to be able to conduct the task in diverse settings, then that is what should be practiced. If a wheelchair user has more than one wheelchair (e.g., powered, manual), because different wheelchairs are used in different settings, he/she should be trained in the use of both.

5.16 Amount of practice

For motor skills to be learned well, they need to be practiced. If a learner is switching from an old to a new coordination pattern, it may take 200 or more practice trials to achieve the change. During the transition, there may be numerous errors, that the learner may find frustrating and discouraging. The amount of practice needed may be much greater (up to 50-fold) for people with injury or disease of the brain.

The "over-learning" strategy (a term that should not be confused with "too much learning") has a positive effect on skill retention. Over-learning means continuing to practice (by 50%–200%) beyond the amount needed for initial success. This can be done right away or during additional practice sessions later. However, more practice is not always better—as the saying goes "Practice does not make perfect, perfect practice does." Also, there may be a point of diminishing returns. More than 4–6 hours of practice a day is unlikely to be productive. If errors begin to occur due to fatigue or frustration, it is probably wise to take a break. For simple tasks, continued practice may actually cause performance to diminish. The literature on wheelchair-skills training suggests that substantial improvements can be made on a group of skills with as little as 2–3 hours of formal training spread over several sessions, but that the target for the clinical setting should probably be higher (e.g., 10–12 hours) if the situation allows. There is no strong evidence as yet regarding the optimum "dose" of wheelchair-skills training.

Although it is not necessary to be an expert to perform a skill in a safe and useful manner, to achieve true expertise at a skill (as a professional athlete, musician, or assembly-line worker may exhibit) may require several hours of practice per day for periods of 10 years or more. There is some evidence to support that millions of repetitions and 10,000 hours of practice may be required for true expertise. Intervals of weeks or months between training are not barriers to learning. As little practice as 6 minutes a month has been shown to be effective. Self-control of the amount of practice and of the practice schedule has been shown to be superior to control by others.

5.17 Facilitating retention

Although a learner may be able to "acquire" a skill during a practice session, it is not uncommon for the learner to fail to perform the skill adequately at the next session. This is a failure of skill "retention." The objective of wheelchair-skills training is long-term retention (i.e., for months and years). For practical purposes, successful performance after such brief intervals as 3 days may need to be accepted as evidence of at least short-term retention, but long-term retention is the goal. The literature on the retention of wheelchair skills is limited but there is evidence to date that skills are retained for periods of a year or more.

There are conditions within and following a practice session that affect whether training on a new skill will be retained. To improve the likelihood of "consolidation," the trainer (and other members of the rehabilitation team) should avoid the introduction of other new skills during the 4–6 hours period following practice. Newly acquired skills may be abolished by subsequent practice of a different novel skill within 4 hours (retrograde interference), especially if the competing task involves the same muscles and movement direction. Similarly, learning one skill can interfere with the subsequent learning of the second skill (antegrade interference). The extent of this interference is related to the duration of the earlier task learning. Performance saturation during training helps consolidation.

Ideally, the learner should sleep before the next training session. Although not always practical, a nap of as little as 40 minutes immediately post-training reduces the susceptibility to interference and results in earlier consolidation. At the subsequent session, the learner may even perform better than at the previous session, without any intervening physical practice. This is sometimes referred to as "off-line learning." Sleep affects some types of skills more than others (sequence-specific skills less so). Sleep is of most benefit to skills that were the most difficult before sleep. Learning by observation and mental imagery is also enhanced by sleep. Anticipated rewards can enhance off-line learning during sleep.

Consolidation begins as a fragile state (one that is susceptible to interference) and progresses over time to a stabilized state. Off-line, a skill becomes less vulnerable to interference (stabilization) and improves in performance (enhancement). During subsequent practice, the consolidated memory can become unstable and susceptible to improvement ("reconsolidation") or deterioration. Older adults have greater susceptibility to interference and fewer off-line gains in motor skills.

5.18 Variability of practice

Most wheelchair skills are of little use if they can only be performed in highly controlled settings. The purpose of wheelchair-skills training is for the learner to use the skill in a variety of settings in his/her life (skill "transfer"). Once a skill is initially acquired and retained, the learner should practice it in different contexts to promote such skill transfer. Diversification may include alterations of the environment (e.g., surface, lighting conditions, time of day, ambient temperature), variations in how the skill is performed (e.g., faster, slower, while multi-tasking) or variations in the learner's state (e.g., with fatigue, anxiety, altered focus of attention). Expanding the scope of training to include a few or many skills in combination (e.g., moving turns on soft surfaces) or in sequence (as might occur while playing a game or going on a community outing) can be very helpful.

To enhance skill retention and transfer, random practice of a group of skills that have already been acquired is generally better than consistent ("blocked") practice, especially for open versus closed skills. However, there will be more errors during random practice. The two approaches are not mutually exclusive. For instance, it may be reasonable to begin with consistent practice and to progress to serial practice of a few skills followed by random practice of those skills. The approach may vary depending upon the personal characteristics of the learner (e.g., children and the elderly do better with less variability and fewer distractions).

The WSTP approach is to make sure that the learner can do each of the basic skills in at least one of the safe and effective methods available (e.g., turning around a corner by pushing harder on the outside hand-rim). To help with skill retention and transfer, trying suitable variations (e.g., adding the drag turn to the learner's repertoire) is encouraged, as well as using the skill in combinations (e.g., a drag turn at the bottom of a ramp). Games can be used to help the learner use the skills in a more automatic fashion, as he/she focuses on the outcome of the game rather than on performing the individual skill.

After maximizing the ability of wheelchair users and caregivers to perform the representative set of individual skills that make up the WSP, toward the end of training these skills can be combined in the various combinations and permutations that make up real life. The WSP skills are the building blocks whereas the real-life activities are the structures that can be built with these units. As part of any such community outings, the learner should be encouraged to plan the route that will be taken—the shortest route is not necessarily the easiest.

Real-life activities provide opportunities to identify challenges requiring intervention and opportunities to learn wheelchair skills as the challenges are encountered ("teachable moments"). However, the order in which such real-life challenges occur is random and inconsistent with a more structured approach in which the sequence of skills learned can be helpful.

5.19 Distribution of practice

Practice may be condensed ("massed") or spread over several sessions ("distributed"). In a rehabilitation center, practice may be organized as brief individual and/or group sessions at regular intervals (e.g., 30 minutes, 1–5 times a week for 2–4 weeks). Sessions might include a warm-up, some time on skills already acquired but requiring further practice, a period during which instruction is received on the principal new skill that is the focus of the session, and a cool-down activity. When the learner has demonstrated the ability to do so safely, the trainer should encourage the learner to practice between formal sessions. Whenever feasible, it is recommended that wheelchair-skills training be spread over a series of brief sessions instead of one long one. Brief practice periods are less likely to conflict with other therapy sessions or to fatigue the learners. For wheelchair users who are elderly, who are unfit, or who have a number of co-morbidities, even a brief session can be fatiguing or cause overuse injury.

One alternative model is to conduct training in and around the learner's home. Another model for learners living in the community is to hold periodic group training courses (e.g., for 1–2 hours, weekly, for several weeks). Another alternative is a skill "camp" (e.g., all day for 1–5 days) in a central location or on a circuit basis. The single-training-session format is commonly used for workshops when training trainers. However, the use of such an approach can cause even highly motivated learners to lose focus and become fatigued. In addition to such problems, this approach may lead to poor retention and consolidation.

The research literature suggests that, for the types of skills that wheelchair users and caregivers need, it is generally less effective to carry out a large amount of training in a condensed manner than it is to spread the training out over a longer period that permits rest and consolidation of what has been learned. However, too much time between practice sessions can allow the learning to decay if the skill has not yet been acquired and consolidated. Beyond this, there is little research evidence to suggest that one of the models noted above is vastly superior to another, so the choice of model(s) can be based on local considerations.

5.20 Whole versus part practice

For skills that consist of a sequence of sub-skills, initially it can be helpful to break the skill down into its components (segmented "motor chunks"). For instance, the stationary wheelie skill can be broken down into three phases—take-off (getting onto two wheels), maintaining balance on two wheels, and landing (returning to the condition of having all four wheels on the ground). The goal, of course, is to build up to the point that the whole skill can be practiced as a unit.

There are some variations on this strategy. For instance, the learner can combine whole- and part-skill practice by focusing attention on different aspects of the skill even though performing the entire skill. If the skill is to be segmented, a progressive approach, from start to finish, is generally preferred because it eventually becomes whole-skill practice. However, the order in which the segments are practiced is not critical. "Chunking" is less often useful for the elderly. Chunking may impair motor skill acquisition, if learners could have taken advantage of cues related to an earlier chunk.

5.21 Simplification and progression

For many wheelchair skills, it is possible to begin with a simpler and less difficult version of the skill. Reducing errors during initial practice attempts may encourage a more implicit method of learning. The learner can master the simpler task before progressing to the ultimate skill level that is the goal of training.

For many wheelchair skills, the simpler version may be useful itself, even if the more difficult levels cannot be learned. For instance, getting the wheelchair up a low curb is a useful skill and also a step toward getting up a high curb. Another example is to learn the wheelie skill in a high-rolling-resistance setting before progressing to a low-rolling-resistance one. This strategy for learning the stationary wheelie has the advantage of reducing the amount of forward–backward movement of the rear wheels needed to maintain balance. This reduces demands on the learner's attention. It also eliminates a degree of freedom (forward–backward movement of the rear wheels). Reducing the degrees of freedom is a strategy that has been observed to be used by beginners learning non-wheelchair skills.

Other examples of progression are adding speed to a task, doing the task in a more challenging environment, adding a second task, reducing the amount of assistance provided by an assistant and reducing the proximity of the spotter. Specific examples of simplification and progression can be found later in the training-tips sections for individual skills. Some of these strategies are similar to those used to increase the variability of practice, with the goal of skill transfer.

In many cases, more difficult skills will build on methods learned in performing simpler but similar skills. For instance, the ability to get over a threshold requires most of the techniques needed when later learning to get up a curb. The order of individual skills listed in Table 1.2 reflects this. As noted earlier, this systematic approach may seem to be conceptually incompatible with the community-outings approach whereby the learner and trainer make forays into the community (e.g., to the corner store) and learn about barriers as they are encountered. However, the two approaches can be used in a complementary fashion, using an initial community outing to help identify skills that require further training and to provide motivation, followed by a systematic process to improve upon those skills, followed by additional community outings to provide variety to the training experiences that encourage skill transfer.

Although a learner can perform a wheelchair skill with any safe and effective method, different methods may be more suitable for some individuals or some situations. For instance, for the "turns while moving forwards" skill as performed by a user of a manual wheelchair who propels the wheelchair with two hands, the basic method is to push harder on the hand-rim of the rear wheel on the outside of the turn. However, for the wheelchair user with good arm function and a wall leading to an opening into which the person wishes to turn, the turn can be accomplished more readily, with less reduction in speed and with less demand on the shoulders if the wheelchair user performs a "drag turn." To do so, the wheelchair user drags the arm along the wall to slow the wheelchair on one side and carry out the turn.

5.22 When the caregiver is the learner

A skill that may not be feasible for a wheelchair user to perform alone may be possible with the assistance of a bystander or caregiver. The training can be directed at the wheelchair user, the caregiver, or the two functioning together. The relationship between a wheelchair user and a caregiver is important. The wheelchair user's needs and preferences should take precedence whenever possible. The wheelchair user may need some help in learning how to ask for help, how to direct the nature of any assistance, and how to politely decline unwanted offers of help.

There are some general considerations when caregivers are the learners. There are ways for caregivers to relate well to wheelchair users. For instance, the caregiver should be instructed to seek permission before taking any actions, to speak clearly, to address the wheelchair user from the front and at eye level whenever possible, and to consider the wheelchair as an item of the wheelchair user's personal property. The caregiver should be cautioned to avoid applying excessive force to the wheelchair user and to avoid sudden movements. The caregiver should always provide the wheelchair user with cues concerning what he/she intends to do before attempting a skill. When the caregiver is successfully trained, the caregiver can serve as a spotter, so the caregiver should be instructed in how to perform in this capacity. The caregiver may also serve as a motivator and trainer (e.g., during practice by the wheelchair user between formal training sessions with the primary trainer). A caregiver can assist with powered wheelchairs in ways similar to manual wheelchairs, even though the powered wheelchair is heavier and bulkier. For instance, with a rear-wheel-drive wheelchair, a caregiver can push down on the back of the wheelchair to unload the casters or to add traction to spinning drive wheels. The caregiver can push a powered wheelchair forward, to assist with overcoming resistance. In addition to these general points, caregiver issues related to specific skills are dealt with later, when those skills are discussed.

Chapter 6 Introduction to safety issues

6.1 General

Wheelchair use can be dangerous. Each year, 5%–18% of wheelchair users experience injuries related to wheelchair use. Of the injuries that are of at least moderate severity, about two-thirds are related to tip-over accidents and/or falling from the wheelchair. Wheelchair users and caregivers are also at risk of chronic injuries, for instance due to poor ergonomic technique.

Improving a person's wheelchair skills may not reduce the likelihood of injury. Providing people with an appropriately setup wheelchair and helping them acquire the abilities and confidence that they need to get around in their communities may, counterintuitively, increase the risk of a tip or collision.

Nevertheless, the goal of wheelchair skills training is for the learner to be able to perform skills safely, effectively, and efficiently. Safety includes both the safety of the wheelchair user as well as of others. If there are two or more ways for a learner to perform a skill and one is considerably safer than the other, the trainer should encourage the learner to use the safer technique. For some learners and some skills that cannot be performed in a consistently safe manner, the most successful outcome of training will be if the learner recognizes that the skill should not be attempted without assistance. A probationary period of supervision may be appropriate before deciding that a person is acceptably safe to use a wheelchair independently.

Because WSP participants are assessed and trained in wheelchair skills with which they may be unfamiliar, participation in assessment and training activities can be dangerous. This section deals with issues affecting safety during these activities. The focus is on the types of risks that can occur and how the spotter can minimize them without unduly interfering with the activity.

Although the safety of WSP personnel (i.e., spotters, testers, trainers) and bystanders is also a concern, this section primarily addresses the safety of the wheelchair user. Although there are a wide range of safety concerns associated with wheelchair use (e.g., hand scrapes, overuse injuries), this section deals only with the major acute risks that a spotter might reasonably be expected to address (e.g., wheelchair tips, falls from the wheelchair).

The best way to spot a skill may vary, depending upon the spotter, the wheelchair user, the wheelchair, and the setting. The material provided in this book, although based on extensive experience with WSP activities, represents only the consensus opinions of the WSP developers. There is no scientific evidence as yet on the best way to spot wheelchair skills.

6.2 What is a spotter?

A spotter is a person who acts to reduce the likelihood of injury to another person who is performing an activity, without unnecessarily interfering with the performance of that activity.

6.3 Who can function as a spotter?

The spotter may be a member of the WSP personnel. Spotter skills can also be useful as wheelchair users go about their everyday activities with friends, family members, and caregivers. Wheelchair users may need to instruct a bystander or passerby on how to best spot a skill that the wheelchair user finds difficult or hazardous.

6.4 Equipment and supplies for the spotter

Spotter strap

A spotter strap is used to assist the spotter in controlling a manual wheelchair during skills during which there is the risk of a rear tip or of the wheelchair rolling away (e.g., down an incline). The necessary requirements of a spotter strap are a means of attaching one end of the strap to the wheelchair, a loop or handle for the spotter's hand at the other end, and sufficient tensile strength to withstand high loads (equivalent to 200 kg or more). One design for such a strap (Figure 6.1) can be found at http://www.wheelchairskillsprogram.ca/eng/spotters.php. Alternatives (e.g., a piece of rope, a dog leash) are equally acceptable if they meet the above-mentioned criteria.

For a wheelchair with an X-shaped cross-brace, the spotter strap is attached where the brace members intersect (Figure 6.2), to avoid any lateral movement of the strap. The low attachment point of the spotter strap helps to resist forward movement of the rear wheels ("submarining") during a rear tip. For a rigid-frame wheelchair, the spotter strap is placed near the midline of a lower frame member

54

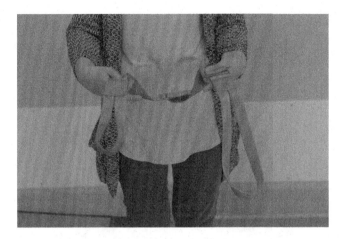

Figure 6.1 Spotter strap.

or camber tube (Figure 6.3), but additional means (e.g., tape) may be needed to keep the strap from sliding sideways. If a knapsack is present or if there are other wheelchair parts (e.g., to provide rigidity to the backrest, to allow the backrest to be folded forward), the path of the spotter strap should be as close to the backrest as possible.

The length of the spotter strap should be adjusted so that it is long enough to allow the spotter to stand upright with the elbow flexed 30°–60° from full

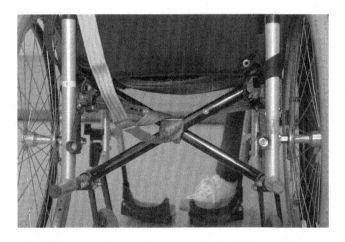

Figure 6.2 Spotter strap with the large loop around the cross-brace of a folding wheelchair.

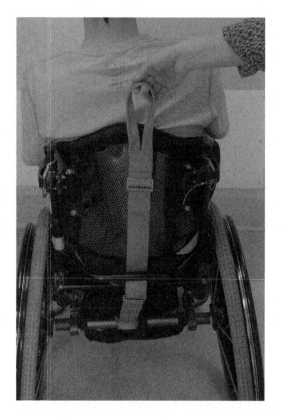

Figure 6.3 Spotter strap attached to the camber tube of a rigid-frame wheelchair.

extension for most skills but short enough that the spotter can be sufficiently close to the wheelchair to intervene. The spotter should hold the hand loop or handle with the palm up and the loop or handle across the palm at the base of the fingers (Figure 6.4), not just across the fingers themselves or around the wrist. The spotter strap should be held ready, but without tension in the strap, because tension can affect the performance of some skills. When not in use, the hand loop or handle can be hung out of the way over a push-handle or other wheelchair part.

Seat belt

For any skills during which there is a risk of the subject pitching or sliding forward out of the wheelchair, a seat belt is recommended. If the subject's wheelchair is not equipped with one, one may be provided by the WSP personnel for training.

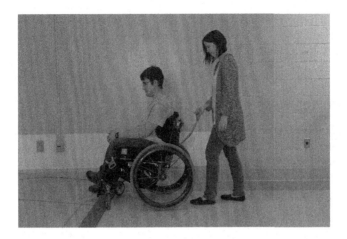

Figure 6.4 Spotter correctly positioned behind a wheelchair.

The wheelchair user may decline to use the seat belt, after being instructed in its availability. A seat belt should not be provided by WSP personnel for WST testing purposes if the subject does not ordinarily use one because this would alter the natural state that is being assessed.

Injury preparedness

A first aid kit should be available, in the event that an injury occurs. Also, a means of communication should be available in the event that the WSP personnel require assistance to deal with an injury.

6.5 Obtaining the subject's permission to be spotted

Wheelchair users with advanced skills perform most of the skills in their daily lives without spotters. Such users may be offended by being spotted unnecessarily. Also, they may be legitimately concerned that inappropriate intervention by a spotter could interfere with the performance of a skill, thereby causing injury.

However, during the initial WST assessment, a spotter is mandatory, at least to the extent of the spotter positioning him/herself where he/she could intervene, if necessary. During subsequent WSP activities, the tester or trainer may permit the subject to waive the spotter, if the tester or trainer is convinced that the subject will not be placed at undue risk by making this decision. It is the subject's right to refuse to be spotted. Indeed, to spot without the subject's permission could be considered a form of assault. However, if the WSP personnel believe that the subject's decision to waive a spotter is inappropriate, the personnel should not permit the subject to participate in WSP activities.

6.6 Spotter warnings to subject

The spotter should let the wheelchair occupant know whenever he/she is or is not in place—the phrases "spotter on" and "spotter off" (with a corresponding pat on the shoulder if in a noisy environment) are useful shorthand means of communicating this information, having explained to the subject what the phrases mean on the first occasion that they are used.

6.7 Ensuring safety during WSP activities

A spotter should be present for any formal WSP activities. The tester or trainer should not permit the subject to attempt or complete any task that he/she has reason to believe that the subject will be unable to complete without risk. For some skills (specified later in the chapter on individual skills), the tester or trainer should ask the subject about whether the subject feels able to perform the skill. For such skills, if the subject believes that he/she would be able to perform the skill, the tester or trainer should then inquire about the intended method to be used. If an unsafe method is described, the tester or trainer should not permit the attempt of that skill in the way described. Despite these precautions, as a general rule, the tester or trainer should try to avoid preemptively disqualifying the subject and should allow him/her to attempt a skill.

Injuries can also occur between skill attempts, while the wheelchair is being moved from one skill site to another or even at rest (e.g., while the spotter steps away to take a phone call). It is the spotter's responsibility to pay close attention to the subject both during and between skill attempts.

6.8 When the spotter should intervene

The spotter should always intervene to prevent a complete tip of the wheelchair, a complete fall from the wheelchair, or a runaway. The spotter should generally not interfere with minimal transient tips (self-limited by definition) that are inadvertent or may even be necessary for the completion of some skills (e.g., getting up a curb). For risks other than tips and falls, it is the WSP personnel's responsibility to stop any skill attempt as soon as it is clear that it is unsafe or about to become unsafe. The WSP personnel should provide feedback to a subject if he/she uses potentially unsafe methods.

6.9 Extent of spotter intervention

The spotter should not intervene to any greater extent than is necessary to ensure that a serious injury is prevented. The extent of spotter intervention may consist of a warning to a subject to stop or change the approach being used, minor physical contact from the spotter (even if the subject is able to complete the trial), or full intervention (e.g., if the subject requires the spotter to prevent him/her from potentially injuring him/herself). If there is significant intervention by WSP personnel during a session, the extent of intervention and the reason for it should be recorded. Note that a spotter may occasionally intervene inappropriately. If this is a minor intervention, that neither hinders nor helps the subject, it can be ignored.

6.10 Stopping a WSP session

If a wheelchair user persists in potentially unsafe activities, despite the warnings of the WSP personnel, the personnel should stop the session and take necessary steps to ensure safety (e.g., contacting the nursing or security staff). This decision will usually be made by the tester or trainer. However, the spotter (if someone other than the tester or trainer) has the right to refuse to participate further, if he/she is concerned about the safety of the subject or personnel.

6.11 Injury determinants

The likelihood and nature of injury varies depending on the wheelchair user and/or caregiver, the wheelchair, and the nature of the skill being attempted. For instance, a wheelchair user who has poor vision, poor judgment, or who is a risk-taker by nature is more likely to be injured than one without these characteristics. Similarly, some wheelchairs are less stable than others. Although this can be an advantage when attempting skills that require the front wheels to be popped off the surface, the trade-off is that such wheelchairs are at a greater risk of an unintentional rear tip.

6.12 Common types of risks and how to prevent them

There are several types of common incidents that can cause injury. Those that require spotter intervention and a general approach to prevent them will be described in this section. Risks during specific individual skills and an approach to preventing them will be described in the chapter on individual skills. Other less acute or less serious injuries (e.g., pinches, scrapes, jarring) are difficult to prevent, because they occur without sufficient time for intervention. These can best be dealt with by training the subject in how to avoid such risks.

Rear tips

A rear tip occurs when the pitch of the wheelchair exceeds the rear stability limit to the extent that the wheelchair user cannot save him/herself and the wheelchair falls backward (Figure 6.5). This may occur while the wheelchair is stationary or moving. If the wheelchair user lets go of the rear wheels during a rear tip, the wheelchair will roll quickly forward while tipping backward. The forward movement is called "submarining."

For most skills that pose a risk of a rear tip, the spotter should be positioned behind the wheelchair with one hand holding a spotter strap (if a manual wheelchair). The spotter may stand in a lunge position (with the forward foot on the opposite side to the hand holding the spotter strap) and close enough to the backrest so that, if the subject tips backward, the spotter can rest the wheelchair on his/her forward thigh for additional support. When using this spotting technique during a skill that requires the spotter to be elevated above the subject or learner (e.g., when descending a curb or incline in the forward direction), the spotter may use a longer spotter strap to reduce any forward bending that could injure the spotter's back. If the spotter catches the subject but cannot return the wheelchair to its upright position, the spotter should inform the subject and then slowly lower the wheelchair backward to the ground. Once the wheelchair is on the ground and the subject is safe and comfortable, the spotter may need to seek additional help to return the wheelchair to the upright position.

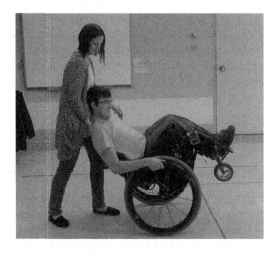

Figure 6.5 Spotter catching wheelchair after a rear tip.

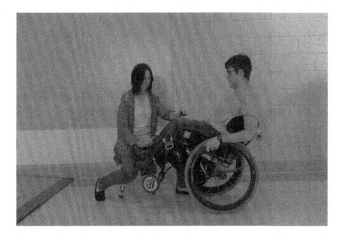

Figure 6.6 Spotter spotting the wheelie skill from in front of the wheelchair.

It is also possible to resist a rear tip from the front (Figure 6.6), for instance when practicing the stationary wheelie skill. The spotter may be positioned just to the side of the front wheels with a hand near the subject's leg or a part of the wheelchair that will not come off if a downward and backward force is applied to it. When a rear tip occurs, the spotter can push down and backward on the leg or wheelchair part to resist the tip and forward movement of the wheelchair.

For many of the skills, the RADs of a manual wheelchair need to be repositioned or removed. While the RADs are inactivated, the WSP personnel need to be particularly attentive to the risk of a rear tip-over. At the end of the session, the WSP personnel should restore the RADs to their original position, unless the subject has progressed to the stage where they can be abandoned.

Forward tips and/or falls

A forward tip occurs when the pitch of the wheelchair exceeds the forward stability limit to the extent that the wheelchair tips forward. This may occur while the wheelchair is stationary or moving. The tip may be partial, but sufficient to allow the wheelchair occupant to slide or fall forward out of the wheelchair. In some instances, such as during a sudden deceleration, the subject or learner may slide or fall forward out of the wheelchair without any tip. When there is a risk of a forward tip/fall and the wheelchair is stationary, the spotter should be positioned in front of and just to one side of the wheelchair. If there is a risk of a forward tip/fall and the wheelchair is moving forward, the spotter may be positioned behind the wheelchair with one hand in front of (but not touching) the wheelchair user's shoulder to prevent a forward tip/fall. However, this can be distracting to the subject and it can be difficult to react quickly enough from this position. A seatbelt can be useful, but

61

there are limitations to its use during some WSP activities. If a second spotter is available, he/she can be positioned to limit the extent of a forward tip or fall.

Sideways tips/falls

A sideways tip occurs when the pitch of the wheelchair exceeds the sideways stability limit to the extent that the wheelchair tips sideways. This may occur while the wheelchair is stationary or moving. The spotter should be positioned to the side to which the tip/fall is expected to occur.

Combination tip/fall risks

Tips and falls do not always occur in the pure rear, forward, or sideways directions. For instance, when descending an incline with one footrest elevated and the other lowered, a combined forward and sideways tip may occur when the lowered footrest strikes the ground at the bottom of the incline-level transition.

Another combination possibility is when different risks present themselves sequentially. For instance, during an attempt to get over a threshold while moving, there is the risk of a rear tip when the wheelchair user attempts to pop the casters high enough to clear the threshold. If the casters do not clear the threshold, the sudden deceleration of the wheelchair can cause a forward tip or fall. When such combination risks are present, the spotter should choose a position where all risks can be minimized. This position will vary, depending upon the skill being attempted and the wheelchair setup. A seat belt or second spotter can be helpful in such situations.

Runaways

A runaway occurs when the wheelchair user loses control of the wheelchair (e.g., when descending an incline) and is unable to bring it to a stop before an injury risk occurs. To prevent the runaway of a manual wheelchair, the spotter should be positioned behind the wheelchair holding a spotter strap. If the wheelchair user loses control, the spotter should pull back on the spotter strap or grasp the push handle to bring the wheelchair to a controlled stop. During the resulting deceleration, the spotter should be alert to the possibility that the subject may fall forward out of the wheelchair, and should position the other hand on the front of the shoulder. A seatbelt or second spotter can be helpful in such situations. Powered wheelchairs or scooters can also runaway if the controls are accidentally activated (e.g., by being caught in loose clothing).

Injury due to contact with a wheelchair part

Pinches can occur when a part of the subject's body gets caught in a wheelchair part (e.g., when opening a folded wheelchair). Injury can also occur if a body part is dragged over or rubbed against a sharp wheelchair part (e.g., the under-surface of a flipped-up footrest). Also, during some activities (e.g., curb ascent) that require the manual wheelchair user to push forcefully on the hand-rims, the backs of the thumbs may get abraded by the wheel locks. During incline descent with a manual wheelchair, the wheelchair user's hands slowing the wheelchair by friction on the hand-rims can experience friction burns or lacerations due to sharp burrs on the hand-rims.

Injuries due to contact with the environment

When the exposed parts of the wheelchair user's body (e.g., hands, feet, head) strike or get pinched by objects in the environment (e.g., doors, walls), injury may occur. The lower limb can be injured if the wheelchair moves forward with the foot planted on the surface. This is most likely to occur when the foot catches on the ground (e.g., at an incline-level transition, when negotiating obstacles or level changes). Examples of injuries are hyper-flexion sprain of the knee or fracture of the tibia or femur due to a knee being forcibly flexed beyond its available range.

Jarring

Sudden jarring forces can be experienced when the wheelchair decelerates suddenly (e.g., when rolling into a threshold, dropping off a curb).

Overexertion injuries

If subjects overexert themselves when attempting skills with which they are unfamiliar or are incapable of performing, they may experience overuse injuries (e.g., affecting the shoulder or back). Similarly, subjects with limited exercise tolerance due to medical conditions (e.g., of heart, lung) may cause themselves harm by overexertion.

Poor ergonomic technique

Subjects are at risk of acute or chronic injuries due to poor ergonomic technique (e.g., folding the wheelchair with a bent and twisted back).

6.13 Dealing with injuries

Despite the best precautions, injuries occasionally occur. Once a tip or fall has occurred, unless this has occurred in a dangerous location (e.g., a city street), there is usually no urgency in getting the wheelchair user back into the upright wheelchair. The WSP personnel can take the necessary time to see whether the wheelchair user has been injured, to assess for wheelchair damage and to formulate a plan. The WSP personnel may need to administer first aid (e.g., cleaning and covering an abrasion). Personnel should have a plan for dealing with any emergency that is beyond their expertise.

6.14 Special considerations when a caregiver is spotted

If a caregiver is the subject, he/she is expected to behave in a manner that is safe for both the wheelchair occupant and him/herself. The spotter in such situations should remain close enough to intervene if the caregiver fails to exercise due caution. A spotter strap held by the spotter is not practical when spotting a caregiver, because this would interfere with the caregiver's performance.

6.15 Special considerations for powered wheelchairs and scooters

For powered wheelchairs and scooters, the spotter's primary strategy is to be in a position where the power can be turned off and, if that fails, to take over the controller (e.g., joystick). For some powered mobility devices, a remote device may be available that allows the WSP personnel or caregiver to intervene by slowing, stopping or steering the wheelchair when a potentially dangerous situation arises. The spotter should also be alert to impending tips or falls. A spotter strap is not a practical solution. A second spotter can be helpful in such situations.

6.16 Risks involved in specific skills

The nature of the skill being attempted should allow the spotter to anticipate the types of injuries that might occur. The chapter on individual skills describes the most common types of risks that should be watched for by the spotter and the usual starting position for the spotter.

Chapter 7 Individual skills

This chapter is organized on the basis of individual skills, in the order listed in Table 1.2. For each skill, the following headings are used:

- *Versions applicable*: For which of the five WSP versions (Table 1.1) this skill is applicable.
- *Skill level*: The skill level for this skill (i.e., basic, intermediate, or advanced).
- *Description*: A brief general description of the skill.
- *Rationale*: The reason why this skill has been included in the WSP skill set.
- *Prerequisites*: If the ability to perform an earlier skill is necessary for this skill to be assessed or trained.
- *Spotter considerations*: If there are other than the general instructions regarding safety discussed earlier, these are mentioned here, in particular the starting position for the spotter and common risks requiring spotter intervention. These considerations are primarily for manual wheelchairs operated by their users but may be adapted for the other versions of the WSP.

Wheelchair Skills Test (WST)

- *WST equipment*: Suggested equipment (other than the wheelchair) and setup (if any) for the WST. Equivalent alternatives may be used. Whenever a "line" is mentioned, it does not need to be visible to the test subject. It may be some other indicator, such as a mark on the floor, a doorway, or a coffee cup on the floor.
- *WST starting positions*: If other than the general starting positions described earlier, the starting positions of the wheelchair user, the wheelchair, and the tester are described. These positions may need to be altered, depending upon the subject's approach to the skill. When a spotter strap is mentioned, this only applies to the version of the WST for manual wheelchairs operated by their users.
- *WST instructions to subject*: An example of the language that the tester might use in directing the completion of the skill. Also, any actions by the tester are noted here. If success on screening questions (*"Can you do it?," "How do you do it?"*) is a strongly recommended precondition to attempting the skill, it is noted here.

- *WST capacity criteria*: The scoring criteria are noted here if there are any beyond the general scoring criteria described earlier. It is also noted here whether failure on a related easier skill may result in an automatic fail without needing to actually attempt the skill.
- *Special considerations*: If the descriptions up to this point for this skill require any special considerations based upon which of the five versions of the WST is being performed (Table 1.1), these are noted here.

Wheelchair Skills Training

- *General training tips*: Tips that apply to most or all of the subsequent sections for this skill.
- *Special considerations*: If the training tips up to this point for this skill require any special considerations based upon which of the five versions of the WSTP is being performed (Table 1.1), these are noted here.

7.1 Moves controller away and back

Versions applicable

Version	Wheelchair Type	Subject Type	Applicable
1	Manual	Wheelchair user	
2		Caregiver	
3	Powered	Wheelchair user	✓
4		Caregiver	✓
5	Scooter	Wheelchair user	✓

Skill level
- Basic.

Description
- The subject moves the controller (e.g., joystick) of a powered wheelchair or scooter away from its usual operating position and then returns it to its original position.

Rationale
- This skill is useful when the controller is in the way for some activities (e.g., approaching a table, feeding, transfers). Some wheelchair users may need to move the controller in order to change the modes or speed. Mounts can vary (e.g., midline flip up, swing away, permanent mounting).

Prerequisites
- None.

Spotter considerations
- Spotter starting position: Beside the wheelchair, on the side of the controller.
- Risks requiring spotter intervention:
  - Runaway.
  - Many units have scissor-like mechanisms that can pinch fingers or clothing.

Wheelchair Skills Test (WST)

Equipment
- None.

Starting positions
- Wheelchair: Controller in its usual operating position (Figure 7.1) and the power off.

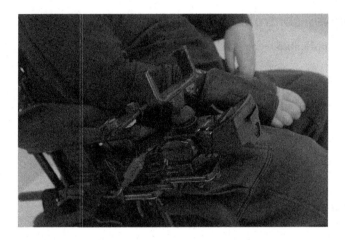

Figure 7.1 Controller in normal operating position.

Instructions to subject
- *"Move the controller out of the way."*
- *"Return the controller to its usual position."*

Capacity criteria
- As for the general scoring criteria, with the clarifications below.
- A "pass" should be awarded if:
  o Within the limits of the controller design, the controller is moved sufficiently out of the way so that it would not interfere with closely approaching a table of the same height as the controller (Figure 7.2).

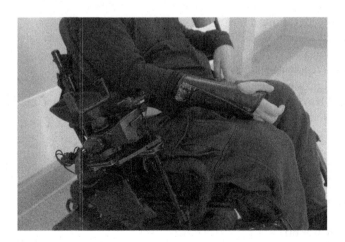

Figure 7.2 Controller in moved-away position.

  o On restoring the controller to the operating position, it should be secured in this position, to the extent possible.
- A "not possible" score can be awarded for this skill because some wheelchairs do not have this feature.

Special considerations for manual wheelchairs operated by users (Version 1)
- Not applicable.

Special considerations for manual wheelchairs operated by caregivers (Version 2)
- Not applicable.

Special considerations for powered wheelchairs operated by users (Version 3)
- None.

Special considerations for powered wheelchairs operated by caregivers (Version 4)
- None.

Special considerations for scooters operated by users (Version 5)
- The controller for a scooter is usually in the midline, on top of the tiller, between the two handles. For many scooters, the tiller can be unlatched and tilted toward or away from the user, to ease transferring onto and off of the scooter.

Wheelchair Skills Training
General training tips

- *Adjustment tips*
  o When attempting to initiate the move-away portion of the skill, it is usually necessary to overcome some initial resistance. The amount of force needed can sometimes be adjusted.
  o Adding a loop to the controller may allow users with limited hand function to independently move the controller.
- The controller should be moved sufficiently out of the way that it would not interfere with approach to a table or to another surface during a transfer.

- *Progression*
  o To avoid runaway, the power should be turned off while this skill is initially being practiced.
  o Training should begin with moving the controller away, then moving the controller back.
  o The skill should eventually be used functionally, such as when approaching a table.

Special considerations for manual wheelchairs operated by users (Version 1)
- Not applicable.

Special considerations for manual wheelchairs operated by caregivers (Version 2)
- Not applicable.

Special considerations for powered wheelchairs operated by users (Version 3)
- The force applied to the controller may need to be applied in a specific location and direction. This location can be identified in such a way that it can be better seen (e.g., with a piece of colored tape) or felt (e.g., with a piece of velcro).
- When moving the controller out of the way, it should not be placed in a position that would make it impossible for the wheelchair user to restore it to its original position.
- If the controller changes its orientation (e.g., by 90°) when it is moved out of the way, the wheelchair user needs to take this into consideration if activating the joystick in this position to avoid driving in an unintended direction.

- *Variations*
  - If the wheelchair user has poor hand control, he/she can use a large, gross motor movement to move the controller. Using the side of the arm or hand along with shoulder movement may allow the controller to be moved independently.
  - The powered wheelchair can be slowly driven at an angle against a fixed external object (e.g., a desk top) to help indirectly push the controller out of the way.

Special considerations for powered wheelchairs operated by caregivers (Version 4)
- None.

Special considerations for scooters operated by users (Version 5)

- The controller for a scooter is usually in the midline, on top of the tiller, between the two handles.
- For many scooters, the tiller can be unlatched and tilted toward or away from the user, to ease transferring onto and off of the scooter.

7.2 Turns power on and off

Versions applicable

Version	Wheelchair Type	Subject Type	Applicable
1	Manual	Wheelchair user	
2		Caregiver	
3	Powered	Wheelchair user	✓
4		Caregiver	✓
5	Scooter	Wheelchair user	✓

Skill level
- Basic.

Description
- The subject turns the power of a powered wheelchair or scooter on and off.

Rationale
- The functions of the powered wheelchair require power. However, when the wheelchair is not being used for position changes or mobility, the power should be turned off when sitting in the wheelchair doing other activities. Otherwise, an article of clothing (e.g., the cuff of a sleeve) can catch on the joystick and unintentionally drive the wheelchair into a person or object. Turning the power off also better maintains the battery charge.

Prerequisites
- None.

Spotter considerations
- Spotter starting position: Beside the wheelchair, on the side of the controller.
- Risks requiring spotter intervention: Runaway.

Wheelchair Skills Test (WST)

Equipment
- None.

Starting positions

- Wheelchair: Power on or off, whichever is the case when the skill assessment begins.
- Scooter: Key in the ignition (Figure 7.3).

71

Figure 7.3 Switch for turning the power on for a scooter.

Instructions to subject
- The order is not important as long as both actions are assessed.
- *"Turn the power on."*
- *"Turn the power off."*

Capacity criteria
- As for the general scoring criteria.

Special considerations for manual wheelchairs operated by users (Version 1)
- Not applicable.

Special considerations for manual wheelchairs operated by caregivers (Version 2)
- Not applicable.

Special considerations for powered wheelchairs operated by users (Version 3)
- None.

Special considerations for powered wheelchairs operated by caregivers (Version 4)
- None.

Special considerations for scooters operated by users (Version 5)
- There is no need for the scooter user to remove and replace the key in the ignition.

Wheelchair Skills Training
General training tips

- *Adjustment tips*
  - A longer lever for the on/off switch will reduce the force required but increase the arc through which the lever must be moved.

o   The location of the on/off switch can vary greatly and may have an impact on independence.
o   Alternative switches can be used for on/off functions (e.g., toggle, depression switch, auxiliary switch).
o   Alternative locations (e.g., head, foot, thigh) can be used for the on/off switch to improve access.

Special considerations for manual wheelchairs operated by users (Version 1)
• Not applicable.

Special considerations for manual wheelchairs operated by caregivers (Version 2)
• Not applicable.

Special considerations for powered wheelchairs operated by users (Version 3)
• The joystick should be in a neutral position before the controller is turned on.
• Turning the controller off while the wheelchair is being operated will bring it to a sudden stop. This can be useful when a sudden stop is needed or if the wheelchair begins to behave erratically.

• *Variations*
o   Rolling the hand onto and off the on/off switch may reduce the need for fine finger dexterity.
o   Using larger movements and body parts may allow users to switch toggle levers on and off independently, if fine motor control is not available.

Special considerations for powered wheelchairs operated by caregivers (Version 4)
• On/off switches may be located on an attendant control unit that can be attached to the wheelchair or operated remotely. Depending on the control method used by the wheelchair user, it may be necessary to turn the controller on before the attendant control can be operated.
• The attendant control overrides that of the wheelchair user.

Special considerations for scooters operated by users (Version 5)
• Turning the power on and off is usually done using a key that can be removed.
• Most scooter users have good upper-limb function. However, for those who do not, the key can be built up to make it easier to grasp and turn.
• Many scooter users leave the key in its receptacle when the power is off. However, to lessen the likelihood of theft when the scooter is left alone, the scooter user may wish to remove the key. If so, removing the key and reinserting it should be practiced.

7.3 Selects drive modes and speeds

Versions applicable

Version	Wheelchair Type	Subject Type	Applicable
1	Manual	Wheelchair user	
2		Caregiver	
3	Powered	Wheelchair user	✓
4		Caregiver	✓
5	Scooter	Wheelchair user	✓

Skill level

• Intermediate.

Description

• The subject operates the controller of a powered wheelchair or scooter to switch between drive modes and speeds and then returns to the original setting.

Rationale

• Most powered wheelchairs (Figure 7.4) and some scooters (Figure 7.5) provide an opportunity for the user to operate the wheelchair in different modes and speeds. The controller settings that are most appropriate for driving slowly in tight quarters are different from the settings that would work best when driving longer distances outdoors or when ascending low curbs.

Figure 7.4 Powered wheelchair user using his thumb to switch the speed setting.

Figure 7.5 Scooter user switching modes.

Prerequisites
• None.

Spotter considerations
• Spotter starting position: Beside the wheelchair, on the side of the controller.
• Risks requiring spotter intervention: Runaway.

Wheelchair Skills Test (WST)

Equipment
• None.

Starting positions
• Wheelchair: Controller in operating position and turned on.

Instructions to subject
• *"Put the wheelchair controller into each of the drive and speed settings that you can, one at a time."*
• *"Put your chair back into the original drive mode/speed."*
• For wheelchairs that have separate controls for the mode and speed settings, if the subject demonstrates one but not the other, he/she may be prompted without penalty (e.g., *"Are there any other ways to adjust the speed or power of the wheelchair?"*).

Capacity criteria
• As for the general scoring criteria, with the clarifications below.
• A "pass" should be awarded if the wheelchair has both adjustable modes and speeds, the subject must be able to handle both for a pass.

- A "not possible" score can be awarded for this skill because all wheelchairs do not have this capability.
- Note that some controllers may appear to have capabilities that the wheelchair cannot perform. For instance, there may be a button labeled "tilt" even for a wheelchair that does not have a tilt mechanism.

Special considerations for manual wheelchairs operated by users (Version 1)
- Not applicable.

Special considerations for manual wheelchairs operated by caregivers (Version 2)
- Not applicable.

Special considerations for powered wheelchairs operated by users (Version 3)
- None.

Special considerations for powered wheelchairs operated by caregivers (Version 4)
- None.

Special considerations for scooters operated by users (Version 5)
- Most scooters have some form of speed control on the tiller (e.g., in the form of a dial) (Figure 7.6), in addition to the lever mechanism that provides moment-by-moment speed control.
- Some scooters have different modes or programs for different operating conditions.
- If the scooter has other operating features (e.g., horn, turn indicators, lights) that are controlled on the "dashboard" of the tiller, the scooter user does not need to be able to operate them to receive a "pass" score.

Figure 7.6 Scooter user using dial to adjust speed setting.

Wheelchair Skills Training

General training tips

- *Adjustment tips*
  - o The type of mode switch used will have an impact on success for some users.
  - o In some wheelchairs, the mode and speed controls are separate.
  - o A controller with the easiest access will be most appropriate for people with cognitive or physical limitations (e.g., three vs. five drive modes, toggle vs. dial for speed control).
  - o Although the manufacturer may provide a representative set of mode settings, the dealer and/or therapist may adjust the settings with a programmer to make them as ideal as possible for the user. These settings can be altered later, as skill improves. For many powered wheelchairs, it is possible to independently select the maximum speed, acceleration, and deceleration in different directions as well as the sensitivity to joystick deflections.
  - o The order of drive modes (e.g., 1, 2, 3, 4) may be different from one wheelchair to the next. For instance, some users may prefer to have the order reflect progressively increasing speed, whereas other users may wish to order the modes to those from the most often to the least often used. Through programming, the dealer and/or therapist can reduce the number of steps to get to the most commonly used drive modes or speeds.
  - o The wheelchair user should be able to see or hear an indication of the mode and speed status.
  - o The process of changing modes may be quite specific. For instance, a switch may need to be activated to make mode selection available, followed by movement of the joystick to the right to move from one mode to the next, followed by movement of the joystick forward to select or use that mode.

Special considerations for manual wheelchairs operated by users (Version 1)
- Not applicable.

Special considerations for manual wheelchairs operated by caregivers (Version 2)
- Not applicable.

Special considerations for powered wheelchairs operated by users (Version 3)
- The user should be trained to select different mode and speed settings for different skills.

- *Progression*
  - o If the powered wheelchair has other operating features (e.g., horn, turn indicators, lights), the trainer should make sure that the user can operate them.

Special considerations for powered wheelchairs operated by caregivers (Version 4)
- None.

Special considerations for scooters operated by users (Version 5)
- Commonly, faster speeds are possible by turning the speed dial clockwise and slower speeds by turning the dial counter-clockwise. These may be graphically illustrated (e.g., with a turtle on the left and a rabbit on the right) (Figure 7.5).
- If the scooter has other operating features (e.g., horn, turn indicators, lights) that are controlled on the "dashboard" of the tiller, the trainer should make sure that the user can operate them.

7.4 Disengages and engages motors

Versions applicable

Version	Wheelchair Type	Subject Type	Applicable
1	Manual	Wheelchair user	
2		Caregiver	
3	Powered	Wheelchair user	✓
4		Caregiver	✓
5	Scooter	Wheelchair user	✓

Skill level

• Basic.

Description

• The subject disengages and engages the motors of a powered wheelchair or scooter.

Rationale

• Disengaging the motors (Figure 7.7) allows the wheelchair to be pushed manually without power (e.g., if the battery is dead). This skill may not be feasible for some wheelchair users due to the characteristics of the wheelchair user and/or the wheelchair.

Figure 7.7 Lever allowing one of the motors of a powered wheelchair to be disengaged ("PUSH" setting) or engaged ("DRIVE" setting).

Prerequisites
- None.

Spotter considerations
- Spotter starting position: Beside the wheelchair, on the side that the subject leans toward.
- Risks requiring spotter intervention: Fall from wheelchair if the subject is a wheelchair user or fall if the subject is a caregiver.

Wheelchair Skills Test (WST)

Equipment
- None.

Starting positions
- Wheelchair: The power should be on and the motors engaged.
- Subject: The subject may get out of the wheelchair to perform this task, but no sitting surface other than the floor or ground may be used.

Instructions to subject
- *"Disengage the motors of the wheelchair, so that the wheelchair can be pushed by hand."*
- *"Engage the motors."*

Capacity criteria
- As for the general scoring criteria, with the clarifications below.
- A "pass" should be awarded if:
  - The tester can confirm that the motors have been disengaged by checking that the wheelchair can be rolled a short distance by pushing on it (Figure 7.8).

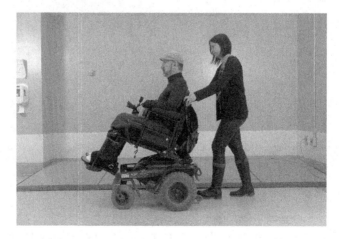

Figure 7.8 Caregiver pushing a powered wheelchair with the motors disengaged.

o The tester can confirm that the motors have been engaged by checking that the wheelchair cannot be rolled.

Special considerations for manual wheelchairs operated by users (Version 1)
- Not applicable.

Special considerations for manual wheelchairs operated by caregivers (Version 2)
- Not applicable.

Special considerations for powered wheelchairs operated by users (Version 3)
- For some powered wheelchairs, the power may need to be turned off for
  the wheelchair to be easily pushed. Failure to do so may result in either
  a "pass with difficulty" or "fail" score depending upon the difficulty that
  the tester experiences in moving the wheelchair.

Special considerations for powered wheelchairs operated by caregivers (Version 4)
- None.

Special considerations for scooters operated by users (Version 5)
- None.

Wheelchair Skills Training

General training tips
- The power should be turned off before the motors are disengaged. The
  wheelchair may be harder to push if the power is on, even if the motors
  are disengaged.
- Depending on the type of wheelchair, rolling the wheelchair slightly
  when disengaging the motors may ease the lever into the disengaged
  position. Some chairs are more difficult than others to push when
  disengaged.
- Various makes and models have different methods of disengaging the
  motors.
- For most powered wheelchairs, there are two motors that need to be
  separately disengaged and engaged.
- Caution should be observed when disengaging the motors on an incline
  because the wheelchair may roll unintentionally and be difficult to stop.

Special considerations for manual wheelchairs operated by users (Version 1)
- Not applicable.

Special considerations for manual wheelchairs operated by caregivers (Version 2)
- Not applicable.

Figure 7.9 Lever allowing the motor of a scooter to be disengaged.

Special considerations for powered wheelchairs operated by users (Version 3)

- It may be possible to perform this task while seated in the wheelchair although it may be necessary to remove the armrests or use a reaching aid.

Special considerations for powered wheelchairs operated by caregivers (Version 4)

- Good ergonomic principles should be used when engaging and disengaging the motors. The caregiver's knees should be bent and the back kept straight. In many cases, a foot can be used to perform the task.

Special considerations for scooters operated by users (Version 5)

- Various makes and models have different methods of disengaging the motor (Figure 7.9).

7.5 Operates battery charger

Versions applicable

Version	Wheelchair Type	Subject Type	Applicable
1	Manual	Wheelchair user	
2		Caregiver	
3	Powered	Wheelchair user	✓
4		Caregiver	✓
5	Scooter	Wheelchair user	✓

Skill level

• Basic.

Description

• The subject operates the battery charger of a powered wheelchair or scooter, setting it up for charging and returning it to the original condition.

Rationale

• Powered wheelchairs and scooters utilize battery power. The battery needs to be charged regularly, as often as daily. The battery charger is usually a separate equipment item (Figure 7.10), often left where the wheelchair is stored overnight. Some are small and light enough to be carried in a knapsack. Some powered wheelchairs have on-board chargers that allow greater flexibility to users when they are working properly but leave the user without a chair if the charger needs to go to the supplier for repairs.

Figure 7.10 Battery charger for a powered wheelchair.

83

Prerequisites
- None.

Spotter considerations
- Spotter starting position: Beside the wheelchair, on the side toward which the subject leans.
- Risks requiring spotter intervention:
- Electrical shock.
- Fall from the wheelchair if the subject is a wheelchair user.
- Fall if the subject is a caregiver.

Wheelchair Skills Test (WST)

Equipment
- The battery charger.

Starting positions
- Wheelchair: Facing the battery charger and at least 0.5 m away from it. The battery charger should be plugged into the power source.
- Subject: The subject may be out of the wheelchair to perform this task and may sit on another surface if there is one available at the charging station used. However, if the subject gets out of the wheelchair, he/she must do so independently.

Instructions to subject
- *"Set up the wheelchair so that the battery can be charged."*
- *"Restore the wheelchair to its original condition."*

Capacity criteria
- As for the general scoring criteria, with the clarifications below.
- A "testing error" should be awarded if the battery charger is not available where the WST is being performed, and the reason for this should be noted in the Comments section.

Special considerations for manual wheelchairs operated by users (Version 1)
- Not applicable.

Special considerations for manual wheelchairs operated by caregivers (Version 2)
- Not applicable.

Special considerations for powered wheelchairs operated by users (Version 3)
- None.

Special considerations for powered wheelchairs operated by caregivers (Version 4)
- None.

Special considerations for scooters operated by users (Version 5)
- For some scooters, the battery may need to be removed from the scooter to be charged.

Wheelchair Skills Training

General training tips

- *Adjustment tips*
  - o If the battery needs to be replaced, a manufacturer-approved model should be used. Failure to do so could cause damage to the battery. The type of battery needed to start an automobile's combustion engine is different from the slow-discharge type needed for a powered wheelchair. Sealed gel batteries are preferable to those with liquid acid that can leak if the battery or wheelchair is tipped over.
  - o If the learner has visual or sensory impairments that affect the orientation of the charger cable and charger port, then bright labels or tactile feedback (e.g., a patch of velcro) can be used to help line up the two components.
  - o The life expectancy of a battery is about 3 years. If a battery does not last a full day of typical use on a full charge, it may need to be replaced.
- The user's manual for the wheelchair may need to be consulted for wheelchair-specific elements of this skill.
- The charger port on the wheelchair is usually near the controller (Figure 7.11) or under the seat. Some wheelchairs have both.
- To avoid electrical shocks, the subject should avoid using the battery charger in a wet environment or where liquids may be spilled on it.

Figure 7.11 Battery charger plug next to receptacle on controller.

- Manufacturers recommend that the battery should not be charged in a room with people present, because there is a risk of explosion with some batteries. This recommendation is difficult to comply with for a wheelchair user acting alone, unless the wheelchair user has a second means of mobility.
- Both the wheelchair and charger should be turned off when being connected to each other and the power source. Then, the power on the charger (if not automatic) should be turned on.
- If the charger cannot be turned off, it is generally better to plug the charger into the wheelchair before plugging it into the wall, to avoid electrical arcing at the charger port.
- The length of time required to charge a battery can vary depending upon the type of charger and the nature of the battery.
- If the battery charger is capable of charging different batteries (e.g., 6 and 12 V), the subject should ensure that the appropriate setting is used.
- A battery with a slightly low charge may function reasonably well on smooth level surfaces but may provide insufficient power to get the wheelchair over obstacles.
- After the battery has been fully charged, it is best to turn the charger off, unplug the charger, and disconnect it from the battery. Although most batteries cannot be overcharged, the life of the charger can be shortened by allowing it to repeatedly activate in response to slight drops in battery charge.
- There is no need to wait for a deep discharge of the battery before recharging.
- It is a good idea to charge a battery on a regular basis, at a frequency that usually prevents the battery from dropping below a 50% charge.
- Chair storage in a climate-controlled environment is best. Extremes of heat or cold are not good for the battery.

Special considerations for manual wheelchairs operated by users (Version 1)
- Not applicable.

Special considerations for manual wheelchairs operated by caregivers (Version 2)
- Not applicable.

Special considerations for powered wheelchairs operated by users (Version 3)
- None.

Figure 7.12 Battery charger on the tiller of a scooter.

Special considerations for powered wheelchairs operated by caregivers (Version 4)

- None.

Special considerations for scooters operated by users (Version 5)

- The charger port on the scooter may be on the tiller (Figure 7.12).

7.6 Rolls forward short distance

Versions applicable

Version	Wheelchair Type	Subject Type	Applicable
1	Manual	Wheelchair user	✓
2		Caregiver	✓
3	Powered	Wheelchair user	✓
4		Caregiver	✓
5	Scooter	Wheelchair user	✓

Skill level
- Basic.

Description
- The subject moves the wheelchair forward a short distance on a smooth level surface.

Rationale
- Forward rolling is a skill used during many wheelchair activities. The short distance is intended to simulate moving about indoors or the distance involved when crossing a two-lane street (Figure 7.13). Most bouts of wheelchair use are relatively short but occur many times a day.

Figure 7.13 Manual wheelchair user crossing a city street.

Prerequisites
- None.

Spotter considerations
- Spotter starting position:
  - If a manual wheelchair, the spotter should be behind the wheelchair, holding onto the spotter strap with one hand.
  - If a powered wheelchair, the spotter should be beside the wheelchair on the side of the controller.
- Risks requiring spotter intervention:
  - If a manual wheelchair, rear tip when accelerating.
  - If a powered wheelchair, runaway, collision.

Wheelchair Skills Test (WST)

Equipment
- A smooth level surface, 1.5 m wide and 10 m long.
- Starting and finishing lines at 0 and 10 m.
- Space at least 1.5 m before the starting line and beyond the finishing line.
- Means of recording time to the nearest second.

Starting positions
- Wheelchair: Stationary, facing the starting line, with the front-wheel axles behind it.

Instructions to subject
- *"Move the wheelchair forward over the finish line without going outside of the boundaries (indicate them)."*
- The tester should indicate where he/she wishes the subject to stop on completion of the skill rather than emphasizing the finish line. Otherwise, the subject may misinterpret the instruction to mean that he/she is supposed to stop just short of the line rather than beyond it. Subjects who stop short of the finish line may be prompted, without penalty, to continue until the axles are over the finish line.

Capacity criteria
- As for the general scoring criteria, with the clarifications below.
- A "pass" should be awarded if:
  - Any safe forward propulsion method is acceptable.
  - The end of the task is when the front-wheel axles cross the finish line and the subject comes to a controlled stop.
  - The wheelchair gently slides along or glances off a lateral barrier as long as there is no injury.
- A "pass with difficulty" may be awarded if:
- The subject takes more than 30 seconds to cover the 10 m distance. Timing this skill provides a means of identifying whether the subject

would be able to get across a street quickly enough to be safe (e.g., when traffic flow is controlled by lights). Although there is considerable variability, most traffic signals provide at least 30 seconds for a full cycle.

- A "fail" score is awarded if a wheel strays outside the lateral boundaries.

Special considerations for manual wheelchairs operated by users (Version 1)

- If the subject strays too close to a wall, it is acceptable for the subject to avoid injuring his/her fingers by pushing off the wall to correct direction.

Special considerations for manual wheelchairs operated by caregivers (Version 2)

- None.

Special considerations for powered wheelchairs operated by users (Version 3)

- None.

Special considerations for powered wheelchairs operated by caregivers (Version 4)

- None.

Special considerations for scooters operated by users (Version 5)

- None.

Wheelchair Skills Training
General training tips

- *Adjustment tips*
  - The distribution of weight on the front and back wheels can be adjusted in some wheelchairs. This has effects on the stability of the wheelchair, traction, and rolling resistance.
  - If the wheelchair user experiences difficulties maintaining a straight direction, the problem may be due to a wheelchair part (e.g., a flat tire) or something rubbing on a wheel (e.g., a seat belt).
- When first attempting to move forward, the direction in which any swivel casters are trailing can lead to some initial resistance to movement or lateral deviation as movement begins. The subject can reposition the casters in the appropriate direction before setting out. Learning how to reposition the casters is a technique that is useful for a number of skills. To reposition the casters, the wheelchair should be moved short distances in a manner that causes the casters to swivel (e.g., forward, then left, then backward, then right).
- The subject should maintain attention in the direction of travel, avoiding distractions to either side but remaining alert to potential hazards.

- Stopping is an important part of this skill. It should be possible to stop the wheelchair at will, on command, and in response to obstacles.

- *Progression*
  - Speed and accuracy are inversely related. It is advisable to begin movement skills with adequate accuracy before increasing the speed.
  - The subject can practice stopping progressively closer to an obstacle, but without touching it. This can include progress from a tall obstacle that can be seen no matter how close the person is to it (e.g., a door), to one that is lost to sight as the user gets closer (e.g., a line on the floor).

- *Variations*
  - The subject can experiment with different speeds.
  - The subject can experiment with how rapidly the wheelchair can be brought to a stop.
  - A strip of bubble wrap can be used for the wheelchair to straddle, providing audible feedback if a straight path is not followed.

Special considerations for manual wheelchairs operated by users (Version 1)
- Each propulsion cycle includes propulsion and recovery phases.

- *Two-hand-propulsion pattern*
  - *Adjustment tip*
    – The wheelchair parts and their setup can affect propulsion. For instance, the rear-wheel axle should be directly under or slightly ahead of the acromion process of the shoulder when the wheel-chair user is sitting upright at rest. The fingers should be able to touch the axle of the rear wheel. When the hands are on the hand-rims of the rear wheels at top dead center, the elbow should not be too straight or bent. These adjustments will allow the wheelchair user to have the hands in contact with the hand-rims in a manner that permits optimal propulsion as described below.
    – The friction between the hands and the hand-rims can be increased by the use of gloves, high-friction covering on the hand-rims, or rubber tubing wrapped in a spiral fashion around the hand-rims.

  - *Propulsion phase*
    – During the propulsion phase, the hands should initially match the speed of the moving wheels to avoid excessive impact loading.
    – The wheelchair user should avoid overly vigorous accelerations that could cause the wheelchair to tip over backward.
    – To propel the wheelchair straight forward, the wheelchair user should grasp the hand-rims and push evenly with both hands. He/she should not wrap the thumbs around the hand-rims, but point them forward.

- The wrists should generally be in a roughly neutral orientation, avoiding the extremes of range.
- To improve friction, if necessary, the wheelchair user can rest the palms of the hands on the tires in addition to using the hand-rims.
- The wheelchair user should lean forward as the elbows are extended, to get more contact time between the hands and the hand-rims and to reduce the chance of a rear tip. This is the first example of a skill that can benefit by leaning. Because the weight of most wheelchair users is large relative to the weight of the wheelchair, leaning can have a major effect on the relative weight on the different wheels. Leaning affects the stability of the wheelchair, traction, and rolling resistance. Leaning is a strategy used often in the later skills.
- To minimize shoulder strain and be mechanically efficient, the wheelchair user should try to push with long, slow strokes, allowing the wheelchair to coast where possible.
- Hand positions can be illustrated by having the wheelchair user imagine the right rear wheel as the face of a clock; the initial (Figure 7.14) and final (Figure 7.15) contact positions for the wheel might then be referred to as 11:00 and 2:00 o'clock. This "three-hour time period" corresponds to a contact angle of 90°.
- To maintain a straight direction during the coast between pushes, the wheelchair user may need to push harder on the side toward which the wheelchair is deviating or use the fingers on the hand-rim to apply friction on the other side. Although it is possible to

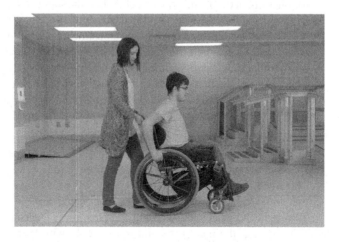

Figure 7.14 For wheelchair propulsion, initial hand contact with the hand-rim at 11:00 o'clock.

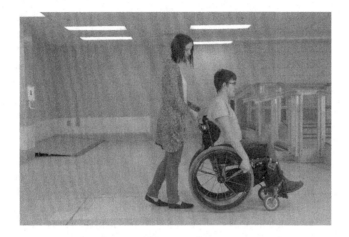

Figure 7.15 For wheelchair propulsion, final hand contact with the hand-rim at 2:00 o'clock.

coast for several meters from a single push, a cadence of about one push per second is commonly used, at least in part to maintain directional control.

o *Recovery phase*
   – A recovery path for the hands below the hand-rims (Figure 7.16) is usually recommended for wheelchair users propelling for any distance on smooth level surfaces.
   – After releasing the hand-rims at the end of the propulsive phase, the arms can be allowed to swing in a relaxed pendular fashion

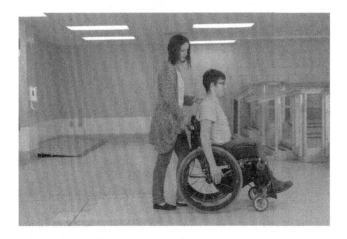

Figure 7.16 For wheelchair propulsion, hand recovery below hand-rim.

below the hand-rims (the "semicircular" recovery pattern) back toward where the propulsive phase will begin for the next propulsive cycle. (The hands need to move slightly outward as well as backward, to avoid contact with the rear wheels.) To reinforce the desired path of the hands, the trainer can ask the wheelchair user to touch the rear-wheel axles during each recovery phase (like a "choo-choo train"). This allows the hands to make initial contact with the hand-rims while moving upward, reducing any impact.
- An additional reason to reach back during the recovery phase and to use long strokes is to exercise the shoulder retractor muscles and maintain shoulder retraction range. This may help to offset the tendency for manual wheelchair users to become round-shouldered due to muscle imbalance and loss of flexibility.
- Wheelchair users with weak or insensitive hands may prefer to slide their hands back along the hand-rims (the "arc" recovery pattern), rather than letting go at the end of the propulsive phase, but any friction will cause some braking to occur. Short strokes with arc recoveries may be appropriate for propelling short distances in confined spaces when fine control is needed.

o *Stopping*
- When stopping, the rate of slowing can be controlled by how hard the hand-rims are gripped. The hand-rims should run through the wheelchair user's hands. While stopping, the hands should be in the 1:00 o'clock position. If the wheelchair user stops too quickly, the wheelchair user may fall forward out of the wheelchair or tip over forward. To prevent this, the wheelchair user should lean back whenever he/she is required to stop quickly.

o *Variations*
- The wheelchair user can see how far he/she can roll on a single push.
- The wheelchair user can see how quickly he/she can cover a distance.
- The wheelchair user can see how quickly he/she can stop on command.
- The wheelchair user can try propelling with one hand at a time.
- The wheelchair user can push an empty wheelchair with one hand, steering with the empty wheelchair.
- The wheelchair user can try propelling with one hand on alternating sides (e.g., as when carrying a cup of coffee).
- The wheelchair user can try to straddle a strip of bubble wrap while coasting, without bursting any bubbles.

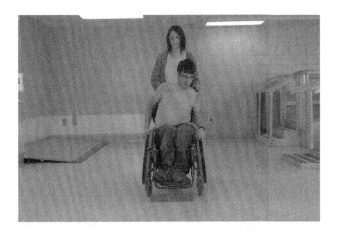

Figure 7.17 Straddling bricks as an exercise to learn about clearance.

- The wheelchair user can try to straddle objects of various heights and widths (e.g., using a few bricks) to better understand the clearance under the wheelchair (Figure 7.17).
- The wheelchair user can pull another occupied wheelchair (with the second wheelchair user holding onto the wheelchair in front) behind him/her (another "choo-choo train" analogy).
- After weaving around objects, it is important to remember to return to the proper propulsion/recovery pattern. An easy, multi-skill activity is to weave through cones and then transition into a few pushes in a straight line before returning to the cones.

- *Hemiplegic-propulsion pattern*
  - *Note*: Hemiplegia due to stroke is used as a representative example of a condition for which foot propulsion can be useful. Wheelchair users with other impairments may find foot propulsion useful as well.

  - *Adjustment tip*
    - The height of the seat should be low enough to allow the full foot to be on the ground when it is directly below the knee.
    - The wheelchair user should wear shoes that do not fall off, that provide protection for the foot, and that provide good traction.

  - *Propulsion phase*
    - If only the sound-side arm is used, the wheelchair will deviate to the weaker side.
    - The wheelchair user propels the wheelchair with the sound-side leg to both propel and steer the wheelchair, with or without the assistance of the sound-side arm.

- There is no need to synchronize the cadence of the hand and foot. Indeed, once moving, some wheelchair users just use the foot to maintain forward movement.
- The propulsion phase for the leg begins with the knee relatively extended (Figure 7.18), pushing down on the floor with the heel, and then flexing the knee under the seat to pull the wheelchair forward (Figure 7.19).

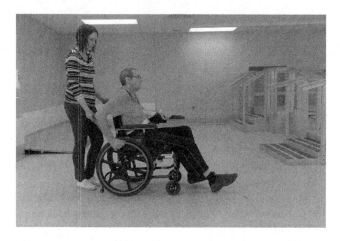

Figure 7.18 Starting position for hand and foot during propulsion phase using the hemiplegic-propulsion pattern.

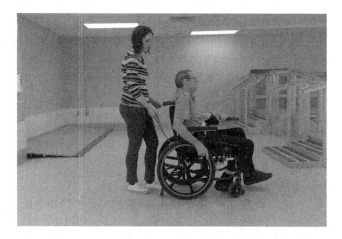

Figure 7.19 Finish position for hand and foot during propulsion phase using the hemiplegic-propulsion pattern.

      – The propulsion phase for the arm is the same as that described above for two hand propulsion.

  o *Recovery phase*
      – At the end of the propulsion phase for the leg, the foot is lifted off the ground, and the knee is extended.
      – The recovery phase for the arm is the same as that described above for two hand propulsion.

  o *Stopping*
      – The wheelchair user may use the hand and foot to stop.

Special considerations for manual wheelchairs operated by caregivers (Version 2)

- If there is only one footrest, because the wheelchair user uses one arm and one leg to self-propel the wheelchair, the unsupported foot can be crossed over the supported one.

Special considerations for powered wheelchairs operated by users (Version 3)

- *Adjustment tips*
  - For this skill and later moving skills, when it is possible to program the wheelchair modes (e.g., with respect to speed, torque, deceleration), the trainer may wish to use a mode that is safest and most likely to be effective when training begins.
  - When set in the slowest speed, there may be a time lag between when a joystick is moved and when the action occurs. This can lead to overcorrection while steering the wheelchair.
  - Nonproportional drives (on/off) are just as dependent on proper programming as proportional drives, if not more so. Setup of nonproportional drives can be graded to include more or less cognitive and physical loads depending on the user's needs and abilities.
  - If the wheelchair user's hand slips off the joystick or control is poor, a different shape for the joystick may be appropriate (e.g., U-shape vs. ball-shape) (Figure 7.20).
  - Powered wheelchairs may be rear-, front-, or mid-wheel drive. The drive configuration will affect the path of the wheelchair and the ease with which the wheelchair can be kept moving in a straight line. For instance, a front-wheel-drive wheelchair tends to be more difficult to keep moving forward in a straight line; some wheelchairs have built-in compensation for this problem.
- This is the first powered wheelchair skill involving movement of the powered wheelchair in a drive mode. With powered wheelchairs, although there are a number of input devices that can be used to control

Figure 7.20 U-shaped attachment for the joystick of a powered wheelchair.

the wheelchair, the term "joystick" has been used because it is the most common device used. Displacing the joystick in a direction will cause the wheelchair to move in that direction. If the controller is of the proportional-control type, the farther the joystick is moved from its rest position, the faster the wheelchair will move in that direction. If the joystick is of the proportional-control type, the user should move it forward gradually to achieve a smooth start. It may take some practice for the wheelchair user to use the joystick in a proportional way—an exercise may be for the wheelchair user to see how slowly he/she can move.

- If the wheelchair user is over-correcting minor deviations from the intended path when driving, changing the contact point with the joystick (e.g., from finger tips to the web-space between the thumb and index fingers) and resting the forearm on the armrest may improve driving smoothness.
- When stopping, the user should allow the joystick to return to the neutral position gradually for a smooth stop. Simply letting go off the joystick will bring the wheelchair to a stop at a rate that has been programmed. Some wheelchairs can be brought to a stop more rapidly if the power is turned off or the joystick is put into reverse.

- *Progression*
  - The subject can practice moving the joystick in an open space and progress to more enclosed ones.
  - The subject can begin at responsive but low torque settings and progress to different modes.

Special considerations for powered wheelchairs operated by caregivers (Version 4)

- When the caregiver is first learning to handle a powered wheelchair, it is preferable to do so with the wheelchair unoccupied, to avoid injury to the wheelchair user.
- Some wheelchairs permit the wheelchair to be operated by a caregiver behind the wheelchair, which is the preferred position.
- For this and other moving skills, the caregiver may operate the wheelchair by using the same joystick that the wheelchair user does. Where space permits, this should be done with the caregiver standing beside the wheelchair and facing forward. In some situations (e.g., going through a narrow opening), the caregiver may need to stand in front of the wheelchair. The caregiver in this situation should be careful not to drive the wheelchair over his/her own feet.
- Standing behind the wheelchair and leaning forward to reach the joystick is not generally recommended.
- Sitting on the wheelchair user's lap to operate the joystick is not generally recommended.

Special considerations for scooters operated by users (Version 5)

- The handles on the tiller control the orientation of the front wheel for steering purposes.
- Lever mechanisms on the handles usually control forward versus backward direction and moment-to-moment speed (Figure 7.21).
- A dial on the tiller controls the general speed setting (high vs. low) depending upon the circumstances.

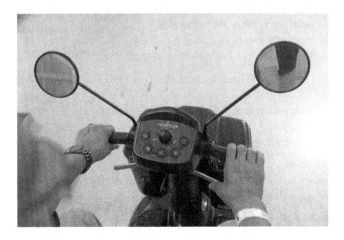

Figure 7.21 Thumb-activated lever mechanism to control forward direction and speed.

7.7 Rolls backward short distance

Versions applicable

Version	Wheelchair Type	Subject Type	Applicable
1	Manual	Wheelchair user	✓
2		Caregiver	✓
3	Powered	Wheelchair user	✓
4		Caregiver	✓
5	Scooter	Wheelchair user	✓

Skill level
- Basic.

Description
- The subject moves the wheelchair backward a short distance on a smooth level surface.

Rationale
- Backward rolling is a skill used during many wheelchair activities. However, a short distance is usually all that is necessary, unless overcoming high-rolling resistance (e.g., on a soft surface, ascending an incline using foot propulsion).

Prerequisites
- None.

Spotter considerations
- Spotter starting position: Behind the wheelchair, holding onto the spotter strap (if a manual wheelchair).
- Risks requiring spotter intervention:
  - o Rear tip when stopping.
  - o Collision with fixed or moving objects.

Wheelchair Skills Test (WST)

Equipment
- A smooth level surface, 1.5 m wide and 2 m long.
- Starting and finishing lines at 0 and 2 m.
- Space at least 1.5 m before the starting line and beyond the finishing line.

Starting positions
- Wheelchair: The back of the wheelchair facing the starting line and the rear-wheel axles behind it.

Instructions to subject
- *"Move the wheelchair backward over the finish line* (indicate it) *without going outside of these boundaries* (indicate them).*"*

- Subjects who stop short of the finish line may be prompted, without penalty, to continue until the rear-wheel axles are over the finish line.

Capacity criteria
- As for the general scoring criteria, with the clarifications below.
- A "pass" should be awarded if:
  - o Any safe backward propulsion method is acceptable.
  - o The end of the task is when the rear-wheel axles cross the finish line and the subject comes to a controlled stop.
  - o A solid barrier is used on either side, the subject may slide along or glance off the barrier.
- Comments only: The subject fails to look backward over the shoulders (Figure 7.22) to monitor that the path is clear. Although this is important in everyday life, in the WST situation, the subject usually backs up into a space that he/she has just moved forward through in the forward direction and has reason to believe is free of obstacles.

Special considerations for manual wheelchairs operated by users (Version 1)
- None.

Special considerations for manual wheelchairs operated by caregivers (Version 2)
- None.

Special considerations for powered wheelchairs operated by users (Version 3)
- None.

Figure 7.22 Shoulder check while rolling backward.

Special considerations for powered wheelchairs operated by caregivers (Version 4)
- None.

Special considerations for scooters operated by users (Version 5)
- None.

Wheelchair Skills Training
General training tips
- If backing up immediately follows rolling forward, then the casters will be trailing backward. As the backing up begins, there may be some initial resistance and directional instability as the casters move into the forward-trailing position. The casters can easily be repositioned by moving them in a circular path.
- The learner should proceed slowly and look over both shoulders regularly to avoid obstacles and collisions.
- Directional stability is more difficult to maintain when backing up a rear-wheel-drive wheelchair. This may lead to a sinuous path, with a series of deviations and over-corrections ("fish-tailing"). This may not be apparent when wheeling backward for a short distance like the 2 m used for the WST, so a longer distance (e.g., 5 m) should be used for training purposes. Slowing down will make it easier for the subject to steer.

- *Variations*
  - Bubble wrap (Figure 7.23) can be place behind a moving rear wheel without the subject's knowledge to provide audible feedback that shoulder checks are needed.

Figure 7.23 Using a strip of bubble wrap to provide feedback on position when rolling backward.

Special considerations for manual wheelchairs operated by users (Version 1)
- In many ways, the technique is the opposite of what is used for rolling forward (as dealt with in the previous skill).
- *Two-hand-propulsion pattern*
  - To propel the wheelchair straight backward, the wheelchair user should reach forward, grasp the hand-rims and pull evenly backward.
  - Some wheelchair users with very weak arms (e.g., people with tetra-plegia) may find it more effective to make contact under the hand-rims with the palms up. Others may prefer to place both hands on the back of the wheels (about 11:00 o'clock, using the clock analogy) with the arms straight and the shoulders shrugged. Then, the wheelchair user can lean back and use the body weight to push down on the wheels.
  - Unlike forward rolling, it is not easy to coast backward without deviating to one side or the other. Therefore, the length of the strokes is usually shorter when rolling backward.
  - Because the distances are usually short, there is no need to use long propulsion strokes or to recover the hands below the hand-rims.
  - To avoid tipping over backward when stopping, the wheelchair user should avoid grabbing the wheels suddenly and should lean forward slightly.

  - *Variations*
    – As for the "rolls forward short distance" skill.
- *Hemiplegic-propulsion pattern*
  - As for the "rolls forward short distance" skill, except the sequence for the leg is to first flex the leg, push down on the floor with the foot enough to ensure good traction, then push the wheelchair backward by straightening the leg.
  - As above for two hand propulsion.

Special considerations for manual wheelchairs operated by caregivers (Version 2)
- The caregiver needs to do regular shoulder checks to avoid collisions or obstacles.

Special considerations for powered wheelchairs operated by users (Version 3)
- *Adjustment tip*
  - The programming of a powered wheelchair is separate for the forward and backward directions. It is possible that a wheelchair that has not been programmed correctly could have difficulty backing up unless the speed control is adjusted upward.
- To move backward, the wheelchair user pulls the joystick backward.
- If the wheelchair is fitted with a rear-view mirror, this lessens the need to turn around to see where the wheelchair is going.

Special considerations for powered wheelchairs operated by caregivers (Version 4)

- None.

Special considerations for scooters operated by users (Version 5)

- As for the "rolls forward short distance" skill, the handles on the tiller control the orientation of the front wheel for steering purposes, lever mechanisms on the handles control forward versus backward direction and moment-to-moment speed, and a dial on the tiller controls the general speed setting (high vs. low) depending upon the circumstances.

7.8 Turns in place

Versions applicable

Version	Wheelchair Type	Subject Type	Applicable
1	Manual	Wheelchair user	✓
2		Caregiver	✓
3	Powered	Wheelchair user	✓
4		Caregiver	✓
5	Scooter	Wheelchair user	✓

Skill level
- Basic.

Description
- The subject turns the wheelchair around to the left and right to face in the opposite direction, while remaining within a confined space.

Rationale
- Turning around in tight spaces is a common challenge for wheelchair users. The type of wheelchair and its dimensions affect the ease with which this skill can be performed.

Prerequisites
- None.

Spotter considerations
- Spotter starting position: Near the wheelchair.
- Risks requiring spotter intervention: No common risks.

Wheelchair Skills Test (WST)

Equipment
- Smooth level surface and a 1.5 m square, marked out by lines on the floor. Solid barriers should not be used unless they are low enough to permit any footrests or anti-tip devices to pass over them.

Starting positions
- Wheelchair: In the center of the square, facing one side of the square (Figure 7.24).

Instructions to subject
- *"Keeping the wheelchair within this square* (indicate it), *turn the wheelchair around until you are facing the opposite direction."*
- *"Now turn the chair in the other direction* (indicate it) *until you are back where you started."*
- It may be helpful if the tester touches the subject's shoulder on the side to which the subject is being asked to turn.

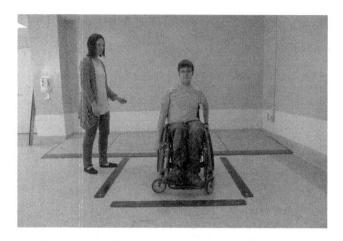

Figure 7.24 Starting position for the "turn in place" skill.

- If the subject has turned, but has not yet turned fully, he/she may be prompted to continue without penalty.

Capacity criteria
- As for the general scoring criteria, with the clarifications below.
- A "pass" should be awarded if:
  - The subject turns at least 160° in each direction (Figure 7.25).
  - Any turning method (e.g., in the wheelie position for a manual wheelchair) (Figure 7.26) is acceptable.

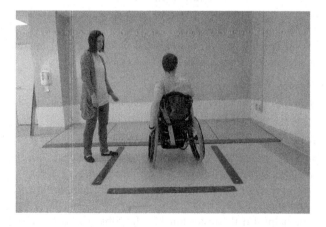

Figure 7.25 Finish position for the "turn in place" skill.

Figure 7.26 Turning in place in the wheelie position.

> o All parts of the wheelchair and subject that touch the ground must remain within the square. However, it is permissible for parts of the wheelchair user's body or wheelchair (e.g., a foot on a footrest) to extend beyond the lines.

Special considerations for manual wheelchairs operated by users (Version 1)
- None.

Special considerations for manual wheelchairs operated by caregivers (Version 2)
- None.

Special considerations for powered wheelchairs operated by users (Version 3)
- None.

Special considerations for powered wheelchairs operated by caregivers (Version 4)
- None.

Special considerations for scooters operated by users (Version 5)
- Because of the way that scooters turn, a three-point turn will usually be necessary to stay within the boundaries.
- Larger outdoor scooters will usually fail this skill.

Wheelchair Skills Training
General training tips

- *Adjustment tips*

- o The ease of making a turn in a tight space depends on the overall length and width of the occupied wheelchair, the distance between the wheels, and how free the casters or steering wheels are to swivel.
  - o The RADs for some wheelchairs increase the overall length of the wheelchairs. Adjusting or removing them may decrease the radius of the turning circle.
- For most wheelchairs (but not scooters), the ability to turn is made possible by casters. Casters are wheels that are free to swivel around a vertical axis. The location of the casters (front vs back) will affect the nature of the turn.
- The footrests of some wheelchairs increase the overall length of the wheelchairs, so a larger turning circle is required. Removing the footrests may make it easier to turn around in close quarters. If the footrests are removed, it is important to avoid injuring the feet by bumping them or running over them with a wheel. If elevated footrests are lowered, the turning circle will be smaller.
- If a wheelchair is in the tilted or reclined position, the turning circle radius may be larger.
- It may be helpful for the learner to shuttle forward and backward—for example, forward turn to the left, backward turn to the right, repeating as necessary—to stay inside the designated space, turning part of the way with each cycle. The longer the chair, the more likely it is that this will be necessary.

- *Progression*
  - o The subject should start with small angular changes of the wheelchair and progress to larger ones.
  - o The subject should start with a larger space in which to turn and progress to smaller ones.
  - o The subject should start at a slow speed, focusing on accuracy (staying within the designated boundaries and progress to faster speeds).

Special considerations for manual wheelchairs operated by users (Version 1)

- *Two-hand-propulsion pattern*
  - o To make the turn more tightly, the wheelchair user should pull back on one wheel, while pushing forward on the other. In such a case, the vertical axis of rotation for the turn is midway between the drive wheels. It may take a few cycles to complete the turn.

  - o *Progression*
    - – The "snap turn" is a more advanced version of the skill. To perform it, the wheelchair user positions one hand well forward and the other well back. Then, in a single uninterrupted motion, the wheelchair user "snaps" the wheelchair around, letting the

hand-rims slide through the fingers until the wheelchair reaches the desired angle. Depending upon the rolling resistance of the surface, the wheelchair may continue to spin in a circle until wheel or hand-rim friction brings the wheelchair to a stop.

- o *Variations*
    - − The skill may be performed in the wheelie position. This minimizes the turning footprint and the corresponding size of the support surface needed, even though the above-ground space needed (i.e., the turning circle) will not diminish to the same extent.
    - − When turning around in confined spaces, it can be helpful for the wheelchair user to push or pull on external objects rather than using the hand-rims.
    - − Game: The learner can be asked to pretend that his/her feet are the hour hand of a clock facing up from the floor and see how quickly and accurately he/she can respond to times that the trainer calls out (e.g., from a starting position of 12:00 o'clock, turn to 3:00 o'clock).

- *Hemiplegic-propulsion pattern*
    - o To turn to the side away from the stronger hand, the wheelchair user should push forward on the hand-rim and push sideways toward the stronger side with the foot.
    - o To turn toward the stronger hand, the wheelchair user should pull back on the hand-rim and push sideways toward the weaker side with the foot.
    - o The wheelchair user may reach across to the opposite wheel with the stronger hand.

Special considerations for manual wheelchairs operated by caregivers (Version 2)
- • To turn in a tight space, the caregiver should pull back on one push-handle, while pushing forward on the other.
- • The caregiver should stand close to the back of the wheelchair if space is limited. If a knapsack prevents this, it can be temporarily removed and placed in the wheelchair user's lap.
- • This skill can be performed in the caregiver-assisted wheelie position.

Special considerations for powered wheelchairs operated by users (Version 3)

- *Adjustment tips*
    - o Adjusting the speed, acceleration, and deceleration for turning will affect the overall turning of the chair.
    - o The location of the drive wheels and seating configuration have impacts on the turning radius of the system.

> o The closer the drive wheels are to the loaded wheelchair's center of gravity, the easier it is to turn in place by simply moving the joystick straight to the left or right. The vertical axis of rotation for such a turn is midway between the drive wheels.

- If the drive wheels are well forward or back, the casters will swing more widely so that a series of to-and-fro motions may be needed to stay within the designated boundaries.

Special considerations for powered wheelchairs operated by caregivers (Version 4)
- None.

Special considerations for scooters operated by users (Version 5)
- Because the drive wheels are not independent and because of the limited angle through which the tiller can turn for most scooters, a scooter cannot turn in place in the same way that manual and powered wheelchairs can.
- The tightness of the turn is also affected by the length of the wheelbase.
- When maneuvering in tight spaces, the speed setting should be reduced.

7.9 Turns while moving forward

Versions applicable

Version	Wheelchair Type	Subject Type	Applicable
1	Manual	Wheelchair user	✓
2		Caregiver	✓
3	Powered	Wheelchair user	✓
4		Caregiver	✓
5	Scooter	Wheelchair user	✓

Skill level
- Basic.

Description
- The subject turns the wheelchair to the left and right while moving forward.

Rationale
- Moving turns are often necessary to avoid obstacles or to change direction.

Prerequisites
- None.

Spotter considerations
- Spotter starting position: Behind the wheelchair, holding onto the spotter strap (if a manual wheelchair), unless the subject has safely performed the "rolls forward short distance" skill, in which case the spotter needs only to be nearby.
- Risks requiring spotter intervention: Rear tip when accelerating.

Wheelchair Skills Test (WST)

Equipment
- 1.5 m-wide level path with a 90° turn. Solid barriers or lines may be used to define the lateral limits.
- Start and finish lines 0.5 m from the corner.
- At least 2 m path before and beyond the corner.

Starting positions
- Wheelchair: Facing the corner, with the front-wheel axles behind the start line.

Instructions to subject
- *"Move the wheelchair forward and turn around this corner* (indicate it)."
- *"Now do the same thing, turning in the other direction."*
- Subjects who stop short of the finish line may be prompted, without penalty, to continue.

Capacity criteria
- As for the general scoring criteria, with the clarifications below.
- A "pass" should be awarded if:
  - The endpoint is when the wheelchair is around the corner, 90° from its original orientation and with the leading wheel axles at least 0.5 m from the corner (Figure 7.27a through c).
  - The subject may touch (or even use) the walls.
  - If lines are used to define the lateral limits, it is permissible for parts of the wheelchair user or wheelchair (e.g., a foot on a footrest) to extend beyond the lines, as long as the wheels or feet on the floor stay within the prescribed limits.

Special considerations for manual wheelchairs operated by users (Version 1)
- None.

Special considerations for manual wheelchairs operated by caregivers (Version 2)
- None.

Special considerations for powered wheelchairs operated by users (Version 3)
- None.

Special considerations for powered wheelchairs operated by caregivers (Version 4)
- None.

Special considerations for scooters operated by users (Version 5)
- None.

Wheelchair Skills Training

General training tips
- The path of the wheelchair parts (e.g., footrests) will differ depending upon the characteristics of the wheelchair (i.e., whether the chair has rear-, mid-, or front-wheel drive). As a general rule when turning, the vertical axis for the turn is midway between the drive wheels, so the farther away from this axis that a wheelchair part or body part is, the greater the circumference through which it will swing.
- When turning around an object (e.g., a corner wall) that the wheelchair is close to, the turn should not begin until the axle of the near-side drive wheel has reached the object.
- When driving a rear-wheel-drive wheelchair toward a 90° turn into a narrow opening, when space is available the wheelchair user should stay as far away as possible from the wall on which the opening is found. This is analogous to parking a car between two other cars in a crowded parking lot.
- If the approach path is narrow but the opening is wide, approaching the corner close to the wall is preferable, watching closely that the axle

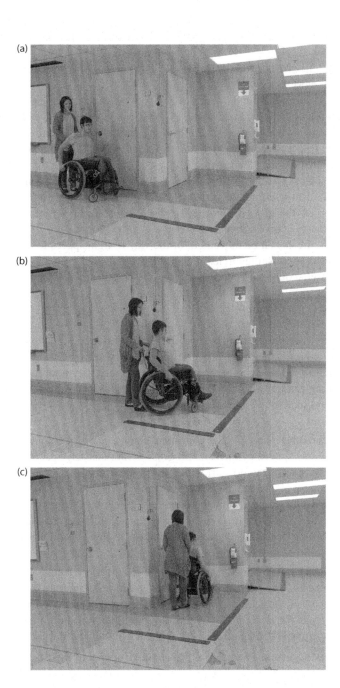

Figure 7.27 For the "turns while moving forward" skill in a manual wheelchair: (a) starting position, (b) mid-turn, and (c) finish position.

of the near-side rear wheel is slightly beyond the corner before turning sharply.

- With a front-wheel-drive wheelchair, there is less of a problem steering a path close to the wall.
- If maneuvering around a series of fixed obstacles that are widely spaced, a useful strategy is to use a path that takes the drive wheels close to the obstacles. If the obstacles are closer together, the wheelchair may need to be driven farther away from each obstacle to have sufficient room in which to complete the turn.
- The subject should clearly understand the difference between the size of the turning circle (i.e., affected by parts, such as footrests, that stick out above the ground) and the size of the turning footprint (that only includes the chair or body parts that touch the ground).
- The footrests can be moved out of the way in tight spaces to reduce the radius of the turning circle.
- The user should be especially careful not to catch the feet on an immovable external object—if the foot stops and the chair continues to turn, a serious injury can result.
- Three-point turns (e.g., using an opening like a doorway to turn around and go back in the opposite direction) can be carried out by making the first turn into the opening while moving forward followed by a backward turn in the opposite direction. Alternatively, the initial turn into the opening can be backward (after rolling past the opening), followed by a forward turn in the opposite direction.

- *Progression*
    - The subject should start with small changes of direction (e.g., around widely spaced pylons) and progress to more closely spaced ones.
    - The subject should start with loose (large-radius) turns and progress to tight (small-radius) ones.
    - When beginning training around full 90° corners, learners may find it easier to break a turn down into its segments—driving straight, turning, then driving straight again, rather than following a smooth curved path.

- *Variations*
    - When using the moving-turns skill in real-life settings, the subject should obey the rules of the road at corners—he/she should slow down if the path around the corner cannot be seen, he/she should stay to the right (if that is the convention in the country in which the training is taking place), and he/she should not cut the corner.

Special considerations for manual wheelchairs operated by users (Version 1)

- *Two-hand-propulsion pattern*
    - When ready to turn, the wheelchair user should slow down the inside wheel and/or push harder on the outside wheel. Slowing down the

inside wheel results in a tighter turn, but causes the wheelchair to slow down. Pushing harder on the outside wheel causes the wheelchair to speed up. The decision on the relative speeds of the two wheels depends on how tight a turn is needed and safety considerations.

o *Variations*
  – The floor space (footprint) needed with all four wheels on the floor is greater than that of a wheelchair in the wheelie position.
  – While coasting in a straight line, the wheelchair user can experiment with the effect that rotating the outstretched arms from side to side has on the wheelchair's direction—swinging the arms to one side causes the wheelchair to turn to the other side.
  – The fixed environment can be used to assist with turning. Timing, intensity, direction, and location of the forces applied to the wall are important features of success. Using the environment minimizes the need to slow down. If the learner is having difficulties, the skill can be simplified by having the trainer push the wheelchair toward the corner while the wheelchair user has the wall-side hand in the ready position and the opposite hand on the lap.
    ⋆ In the "drag" turn, the wheelchair user drags a hand, in a rear position, along the wall to turn toward the wall and around the corner (Figure 7.28).
    ⋆ In the "push-off" turn, the wheelchair user uses a hand, in a forward position, to push away from the wall (Figure 7.29a and b).

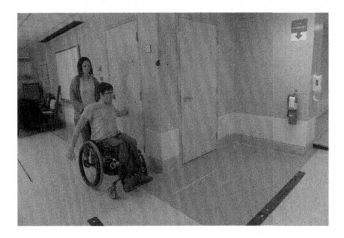

Figure 7.28 Wheelchair user using a rear-placed hand on wall to initiate a "drag turn" to the left.

115

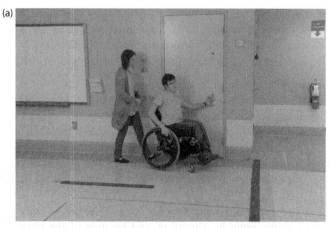

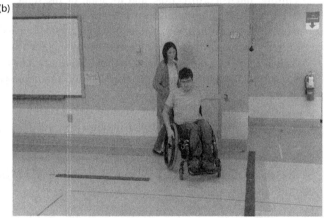

Figure 7.29 Wheelchair user using a forward-placed hand on wall to initiate a "push-off turn" to the right. (a) Starting position and (b) mid-turn.

- *Hemiplegic-propulsion pattern*
  - The wheelchair user should use the foot to help steer.
  - It is easier to turn away from the sound (unaffected) side than toward it.

Special considerations for manual wheelchairs operated by caregivers (Version 2)

- The caregiver should push harder with the push-handle on the outside of the turn and pull back slightly on the inside handle.
- The caregiver should be careful to avoid having the wheelchair user's hands or feet hit any barriers.
- The wheelie position can be used to turn in tight spaces.

Special considerations for powered wheelchairs operated by users (Version 3)

- The path of the wheelchair is affected by whether the wheelchair is rear-, mid-, or front-wheel drive. The general rule of paying attention to the axle of the near-side drive wheel applies.
- If the leading wheels are the drive wheels (i.e., a front-wheel-drive wheelchair), the trailing casters may swing wide of the path and may strike the wall on the far side, depending upon the radius of the turn.
- If the wheelchair is about to collide with the corner, the wheelchair user should not reach out to fend off with the hands or feet—this is ineffective and may cause injury. The body parts should be kept within the protective envelope of the wheelchair.

Special considerations for powered wheelchairs operated by caregivers (Version 4)

- None.

Special considerations for scooters operated by users (Version 5)

- Some scooters have three wheels and some have four. All other things being equal, a three-wheeled scooter will corner better but will be more vulnerable to sideways tips.
- Unlike powered wheelchairs, the drive wheels do not operate independently. Steering the scooter is related to the orientation of the front wheel(s), controlled by the handles of the tiller (Figure 7.30a through c).

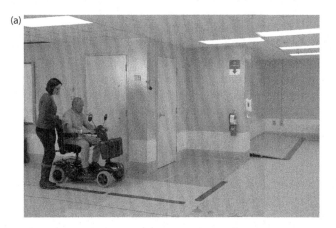

Figure 7.30 For the "turns while moving forward" skill in a scooter: (a) starting position. (*Continued*)

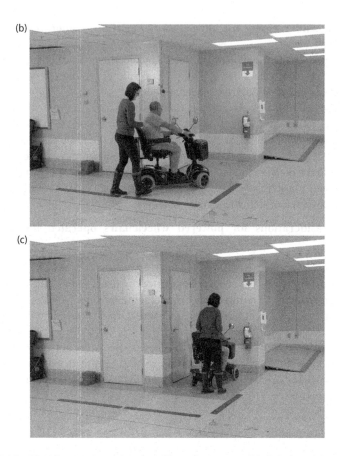

Figure 7.30 (Continued) (b) Using the tiller to steer and (c) finish position.

- Because most scooters are rear-wheel drive, turning is similar to driving a car, an analogy that may be useful.
- Because most scooters have long wheelbases in comparison with other wheelchairs and because there are usually limits to how far the handles can be turned, scooters cannot turn as tightly as other wheelchairs.

7.10 Turns while moving backward

Versions applicable

Version	Wheelchair Type	Subject Type	Applicable
1	Manual	Wheelchair user	✓
2		Caregiver	✓
3	Powered	Wheelchair user	✓
4		Caregiver	✓
5	Scooter	Wheelchair user	✓

Skill level
- Basic.

Description
- The subject turns the wheelchair to the left and right while moving backward.

Rationale
- Moving turns in the backward direction are often necessary to avoid obstacles or to change direction. However, for most wheelchair users, such turns are usually required less often in everyday life than moving turns in the forward direction.

Prerequisites
- None.

Spotter considerations
- Spotter starting position: Behind the wheelchair, holding onto the spotter strap (if a manual wheelchair), unless the subject has safely performed the "rolls backward short distance" skill, in which case the spotter needs only to be nearby.
- Risks requiring spotter intervention:
  - o Rear tip when stopping.
  - o Collision.

Wheelchair Skills Test (WST)

Equipment
- As for the "turns while moving forward" skill.

Starting positions
- Wheelchair: The back of the wheelchair facing the corner, with the rear-wheel axles behind the start line.

Instructions to subject
- *"Move the wheelchair backward and turn around this corner* (indicate it)."

- *"Now do the same thing, turning in the other direction."*
- Subjects who stop short of the finish line may be prompted, without penalty, to continue.

Capacity criteria

- As for the general scoring criteria, with the clarifications below.
- A "pass" should be awarded if:
  - The endpoint is when the wheelchair is around the corner, 90° from its original orientation and with the leading wheel axles at least 0.5 m from the corner.
  - If lines are used to define the lateral limits, it is permissible for parts of the wheelchair user or wheelchair (e.g., a foot on a footrest) to extend beyond the lines, as long as the wheels or feet on the floor stay within the prescribed limits.

Special considerations for manual wheelchairs operated by users (Version 1)

- None.

Special considerations for manual wheelchairs operated by caregivers (Version 2)

- None.

Special considerations for powered wheelchairs operated by users (Version 3)

- None.

Special considerations for powered wheelchairs operated by caregivers (Version 4)

- None.

Special considerations for scooters operated by users (Version 5)

- None.

Wheelchair Skills Training

General training tips

- As for the "turns while moving forward" skill.

Special considerations for manual wheelchairs operated by users (Version 1)

- As the "turns while moving forward" skill.

Special considerations for manual wheelchairs operated by caregivers (Version 2)

- As the "turns while moving forward" skill.

Special considerations for powered wheelchairs operated by users (Version 3)

- Although operation of the joystick is fairly intuitive when performing turns while moving forward (e.g., if one wishes to turn to the right, the joystick is moved to the right), it can be difficult to get used to performing moving turns in the backward direction. It can be helpful to remember that the left–right direction in which the joystick should be displaced should be the direction in which the wheelchair user wishes his/her knees to move. For instance, when making a backward turn to the left, the knees will move to the right, so that is the direction toward which the joystick should be displaced.

Special considerations for powered wheelchairs operated by caregivers (Version 4)
- As for Version 3.

Special considerations for scooters operated by users (Version 5)
- Having a mirror attached to a handle can be useful when driving straight back but is of less use when turning—as one backs up and turns to the right, the mirror looks to the left.

7.11 Maneuvers sideways

Versions applicable

Version	Wheelchair Type	Subject Type	Applicable
1	Manual	Wheelchair user	✓
2		Caregiver	✓
3	Powered	Wheelchair user	✓
4		Caregiver	✓
5	Scooter	Wheelchair user	✓

Skill level
- Basic.

Description
- The subject maneuvers the wheelchair sideways to the left and right.

Rationale
- Repositioning the wheelchair sideways in a tight space is commonly necessary to get closer to or farther away from objects (e.g., a bed, table).

Prerequisites
- None.

Spotter considerations
- Spotter starting position: Near the wheelchair.
- Risks requiring spotter intervention: No common risks.

Wheelchair Skills Test (WST)

Equipment
- Target lateral barrier or line.
- A line 10 cm from the target.
- Means to limit the extent of forward–backward movement to 1.5 m. If these limits are solid barriers, they must be low enough that any footrests or anti-tip devices can pass over them. A strip of bubble wrap can be used to provide audible feedback to the tester that the wheels have gone beyond the limits of the space available.
- Note that the same setup can be used to test sideways maneuvering to left and right by simply turning the wheelchair to face in the opposite direction between the two attempts.
- Alternatively, and more simply, the end position for the first attempt can be the starting position for the second attempt.

Starting positions
- Wheelchair: Parallel to the target with the closest wheel at least 0.5 m from it (Figure 7.31).

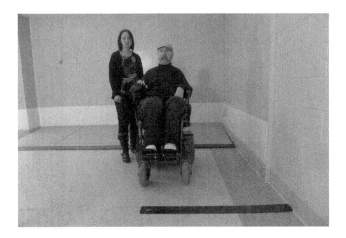

Figure 7.31 Starting position for the "maneuvers sideways" skill.

Instructions to subject
- *"Get this wheel* (indicate the one closest to the target) *as close as you can to this wall/line* (indicate it), *using the space available* (indicate the forward–backward limits)."
- Repeat after getting the other side of the wheelchair into the starting position.
- If the wheelchair is close to the desired finish position, but not quite there (too far away or at too great an angle), it is permissible to prompt the subject without penalty (e.g., *"Can you get a little closer?"*, *"Can you straighten out the wheelchair?"*).

Capacity criteria
- As for the general scoring criteria, with the clarifications below.
- A "pass" should be awarded if:
  - The most lateral aspect of the wheelchair is moved to within 10 cm of the target (Figure 7.32). For manual wheelchairs, the most lateral aspect of the wheelchair will usually be the rear-wheel hand-rim. For powered wheelchairs, this will usually be the drive wheel. The wheelchair may touch the target.
  - On completion, the fore-aft axis of the wheelchair must not be at an angle of greater than 20° from the wall.
  - The parts of the wheelchair or subject in contact with the ground must stay within the 1.5 m forward–backward limits, but other parts of the wheelchair or subject (e.g., feet on footrests) may extend beyond these limits without penalty.
  - It is permissible for the subject to move farther away from the target to allow an approach from the front or back, as long as the wheelchair does not go beyond the 1.5 m separating the front and rear barriers.

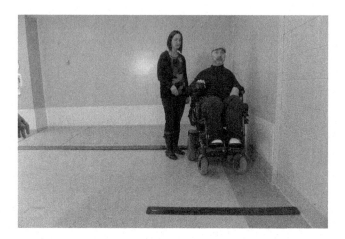

Figure 7.32 Finish position for the "maneuvers sideways" skill.

Special considerations for manual wheelchairs operated by users (Version 1)
- Most subjects will use to-and-fro motions (as in parallel parking a car), but "bunny hopping" or rocking (from the wheels on one side to those on the other) is permitted.

Special considerations for manual wheelchairs operated by caregivers (Version 2)
- The caregiver's feet need to stay within the available space.

Special considerations for powered wheelchairs operated by users (Version 3)
- None.

Special considerations for powered wheelchairs operated by caregivers (Version 4)
- The caregiver's feet need to stay within the available space.

Special considerations for scooters operated by users (Version 5)
- Larger outdoor scooters will usually fail this skill.

Wheelchair Skills Training
General training tips

- *Adjustment tip*
  - The length of the wheelchair can sometimes be minimized through set up (e.g., by moving the axles forward).
- The subject needs to be aware of the widest and longest points of the wheelchair.
- The trainer may use the analogy of parallel parking a car (although without being able to pull forward ahead of the opening as one would do in a car), if the subject has had such experience.

- If the space available is limited, the subject may need to shuttle the wheelchair forward and backward a number of times to get into the desired position, moving more to the side with each attempt.
- *Progression*
  o The subject should start with ample forward–backward room in which to maneuver and gradually decrease the space available. Strips of bubble wrap can be used to provide audible feedback regarding the forward and backward limits.
  o The subject should start with small sideways steps and progress to larger ones.
  o The subject should start at a slow speed, focusing on accuracy (staying within the designated boundaries), increasing the speed within the limits of accuracy.
  o The subject may perform the skill in the wheelie position when that skill has been learned.
- *Variations*
  o The subject may mimic parallel parking a car, pulling forward ahead of the target opening, then backing into the opening.
  o The subject may use the sideways-maneuvering technique to negotiate to the other side of a barrier with a gap in it (e.g., two bolsters in a parking lot) that is too narrow to drive straight through but is low enough from the ground to allow clearance between the wheels. It may be possible to move one pair of wheels through the gap at a time, transiently straddling the obstacles with one pair of wheels on either side of the obstacles and the wheelchair parallel with the obstacles (Figure 7.33a through e). The wheelie position can be very helpful in performing this skill (Figure 7.34).

Special considerations for manual wheelchairs operated by users (Version 1)

- *Two-hand-propulsion pattern*
  o *Variations*
    – An alternative for the wheelchair user with good upper body strength and coordination is to use the "bunny-hop" method. To do so, the wheelchair user hops the rear wheels to the side by shifting the body weight in the desired direction and pulling up on the rear wheels to have them move in the same direction. The wheels do not need to get fully off the ground to be successful. This is most useful when space is very limited. Initially, the wheelchair user can get used to just hopping up and down, with no sideways movement. If the hands holding onto the hand-rims are not at the top dead center, the rear wheels will rotate when they become unloaded. This can be prevented by applying the wheel locks.

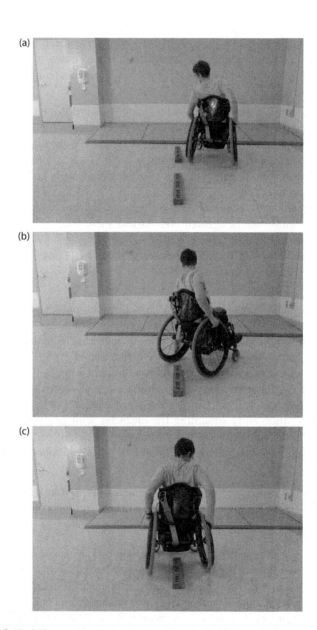

Figure 7.33 Variation on the "maneuvers sideways" skill to get through a narrow opening. (a) Starting position, (b) one wheel passing through the opening, (c) Straddling the obstacles. *(Continued)*

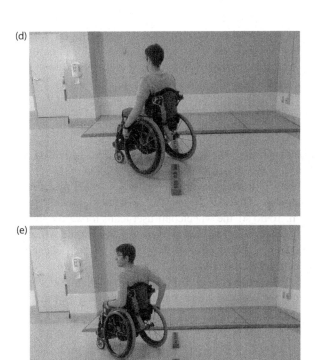

Figure 7.33 (Continued) (d) second wheel passing through opening, and (e) finish position.

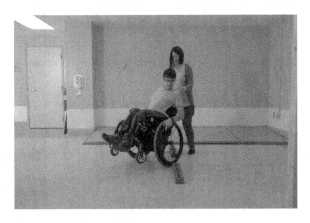

Figure 7.34 Variation on the "maneuvers sideways" skill to get through a narrow opening, using the wheelie position.

    – A similar effect can be created by rocking the wheelchair from side to side. The wheelchair user should lean hard in the direction that he/she wishes to move and return more gently to the upright position.

- *Hemiplegic-propulsion pattern*
  - ○ No special considerations.

Special considerations for manual wheelchairs operated by caregivers (Version 2)

- The caregiver should generally not attempt to lift the occupied wheelchair sideways. However, it may be possible to use the "wheelbarrow" approach. To do so, the wheelchair user leans forward to unload the rear wheels, being careful not to tip over or fall from the wheelchair. Then, the caregiver may be able to slightly lift the rear wheels and move them sideways in small increments.
- The caregiver should be careful that the wheelchair user's arm or hand is not caught between the lateral barrier and the rear wheel.

Special considerations for powered wheelchairs operated by users (Version 3)

- *Adjustment tip*
  - ○ A mirror attached to the wheelchair can be used to provide visual feedback on the position of the chair with respect to the rear barrier.
- The strategies for front-wheel-drive and rear-wheel-drive wheelchairs are somewhat different. For instance, when maneuvering away from a wall that is very close, it is helpful to move the casters away from the wall first.

Special considerations for powered wheelchairs operated by caregivers (Version 4)

- None.

Special considerations for scooters operated by users (Version 5)

- *Adjustment tip*
  - ○ A mirror attached to a tiller handle can be used to provide visual feedback on the position of the chair with respect to the rear barrier.
- Because of the long wheelbase of most scooters, it is often not possible to move sideways when the amount of space is very limited but the skill should still be practiced in larger spaces.

7.12 Reaches high object

Versions applicable

Version	Wheelchair Type	Subject Type	Applicable
1	Manual	Wheelchair user	✓
2		Caregiver	
3	Powered	Wheelchair user	✓
4		Caregiver	
5	Scooter	Wheelchair user	✓

Skill level
- Basic.

Description
- The subject reaches up to touch a high object.

Rationale
- A combination of upward and sideways or forward reaching is often needed when reaching for a light switch, elevator button, or cupboard. This skill is not included in the caregiver skill set because it is not a challenge for most caregivers.

Prerequisites
- None.

Spotter considerations
- Spotter starting position: Near the wheelchair, on the side toward which the subject leans (if any).
- Risks requiring spotter intervention:
  - Forward or sideways tip or fall when reaching, leaning, or standing up.
  - Forward fall or tip due to standing on footrest.

Wheelchair Skills Test (WST)
Equipment
- Target about 2.5 cm in diameter and 1.5 m above the floor (Figure 7.35).

Starting positions
- Wheelchair: Facing the target with the front-wheel axles at least 0.5 m away.

Instructions to subject
- *"Touch the target (indicate it). You may move your wheelchair."*

129

Figure 7.35 For "reaches high object" skill, powered wheelchair user touching mark on the wall. The horn is a fun option for training.

Capacity criteria

- As for the general scoring criteria, with the clarifications below.
- A "pass" should be awarded if:
  o The subject reaches up under control, touches the target and then resumes the normal sitting position.
  o The subject may use either hand.
  o A reaching aid may be used, if it is carried by the subject.
  o The subject chooses to remove or reposition parts of the wheelchair (e.g., the footrests) to improve the reach (e.g., by standing), this is permitted as long as the subject can remove and replace the parts independently. After touching the target, the subject may be prompted, without penalty, to restore the wheelchair to its original state.
  o A wheelchair with a stand-up or seat-elevation feature may be used, as long as the subject can operate it independently.
  o The subject may get out of the wheelchair to perform this skill.
- A "pass with difficulty" score should be awarded if the wheelchair user chooses to stand to accomplish the task and does not lock the wheel locks or clear the footrests away. Not providing a fail score is in recognition that some wheelchair users can accomplish the task in a careful and safe manner without these precautions.
- A "fail" score should be awarded if:
  o A wheelchair user attempts to stand with a foot on a footrest, the spotter should intervene. However, some wheelchairs (e.g., those

with the footrests behind the casters) may allow the skill to be safely performed in this way.

o The wheelchair user stands up without applying the wheel locks and the wheelchair rolls backward far enough to cause a fall.

Special considerations for manual wheelchairs operated by users (Version 1)
• None.

Special considerations for manual wheelchairs operated by caregivers (Version 2)
• Not applicable.

Special considerations for powered wheelchairs operated by users (Version 3)
• None.

Special considerations for powered wheelchairs operated by caregivers (Version 4)
• Not applicable.

Special considerations for scooters operated by users (Version 5)
• None.

Wheelchair Skills Training

General training tips

• *Adjustment tip*
  o Chair height and the overall length of the wheelchair can have impacts on the wheelchair user's ability to reach overhead objects, depending upon the methods used.
• The wheelchair should be positioned to take advantage of the subject's reach, strength, and balance.
• Reaching and leaning reduce stability, putting the wheelchair user at risk of falling out of the wheelchair or, if in a manual wheelchair, tipping the wheelchair over.
• The learner may use a reaching aid, but should carry it with him/her.
• For a person with weak trunk muscles, to avoid falling in the direction that he/she is leaning, he/she should hook the nonreaching arm behind the push-handle or hold onto the armrest or wheel.
• To help right him/herself in the chair after reaching for the object, the wheelchair user can pull on the opposite armrest or wheel.
• If the armrest on the side to which the wheelchair user wishes to reach is moved out of the way, it allows the wheelchair user to bend further sideways.

- The wheelchair user needs to exercise caution when reaching across the body, especially when reaching for or picking up something (e.g., a heavy object on a high shelf, hot coffee, a knife) that could injure the user if it were spilled or dropped onto the lap. Also, bending and twisting at the same time can cause back injury.

- *Variations*
  - If the wheelchair user is reaching for a light and unbreakable object from a high shelf, he/she can use an improvised reaching aid (e.g., a rolled up magazine, a cane) to help move the object off the shelf and catch it. In a store, when an object is out of reach, an object (e.g., a cereal box) on a lower shelf can be used to ease the desired object off the higher shelf so it can be caught.

Special considerations for manual wheelchairs operated by users (Version 1)

- *Adjustment tip*
  - Caster locks can be helpful to keep the casters oriented in the correct direction (trailing in the direction of lean).
- To be safer when leaning or bending forward, the wheelchair user can move the footrests out of the way and place the feet on the floor.
- If standing up, the wheelchair user should first apply the wheel locks and clear the footrests out of the way. If the wheelchair user stands up on the footrests, a forward tip is likely unless the footrests are behind the front wheels. If standing, the wheelchair user should keep one hand on the wheelchair to keep from falling.
- It is sometimes easier to approach the target backward, but the wheelchair user needs to be careful not to reach backward too far and tip the wheelchair over.
- If the wheelchair user chooses to lean forward to accomplish the task, he/she should make sure the casters are trailing forward to decrease the likelihood of tipping forward. When the casters are trailing forward, they lie ahead of the portion of the wheelchair frame to which they are attached, as is the case when the wheelchair is rolled backward. This is another good opportunity to teach the wheelchair user about how to swivel the casters into different directions if it has not been covered earlier in training. Caster swivel control is a skill that will be useful for later skills. To swivel the casters 180° in a tight space requires that a combination of forward–backward and left–right forces be applied to the casters. As an exercise, the trainer can ask the learner to point the casters at targets or to pretend the caster is the hour hand on a clock (*"Set your caster to 3:00 o'clock"*). Alternatively, the trainer can ask the learner to swivel the casters around an object (e.g., a coin) on the floor without touching it.

Special considerations for manual wheelchairs operated by caregivers (Version 2)
- Not applicable.

Special considerations for powered wheelchairs operated by users (Version 3)
- If the wheelchair can be repositioned (e.g., with respect to tilt, recline, seat height), this may be helpful. For instance, if the wheelchair user's balance is good and his/her feet can be placed on the floor, the wheelchair user can move to the front of the seat and obtain help in rising from the tilt mechanism.

Special considerations for powered wheelchairs operated by caregivers (Version 4)
- Not applicable.

Special considerations for scooters operated by users (Version 5)
- When getting out of the scooter, the scooter user should keep at least one hand on the scooter for balance if that is an issue.

7.13 Picks object from floor

Versions applicable

Version	Wheelchair Type	Subject Type	Applicable
1	Manual	Wheelchair user	✓
2		Caregiver	✓
3	Powered	Wheelchair user	✓
4		Caregiver	✓
5	Scooter	Wheelchair user	✓

Skill level
- Basic.

Description
- The subject picks a small object up from the floor.

Rationale
- Objects that need to be picked up from the floor or ground vary from those as small and light as a coin or a piece of paper to those as bulky and heavy as a young child.

Prerequisites
- None.

Spotter considerations
- Spotter starting position: Near the wheelchair, on the side toward which the subject leans (if any).
- Risks requiring spotter intervention:
  - Forward or sideways tip or fall when reaching, leaning, or standing up.
  - Forward fall or tip due to standing on footrest.
  - Rolling the wheels over the fingers.

Wheelchair Skills Test (WST)

Equipment
- Object about the size of a paperback book (dimensions about 5 cm × 10 cm × 10 cm and weighing <0.2 kg) placed flat on the floor. Any object of roughly equivalent size and weight may be used.

Starting positions
- Wheelchair: Facing the target with the front-wheel axles at least 0.5 m away.

Instructions to subject
- "*Pick up the object* (indicate it). *You may move your wheelchair.*"

Capacity criteria
- As for the general scoring criteria, with the clarifications below.
- As for the "reaches a high object" skill except:
  - The finishing position is with the object in the lap or in the hand and the wheelchair user sitting upright. The subject may use either hand.

Special considerations for manual wheelchairs operated by users (Version 1)
- When bending or stooping to pick up the object, the caregiver may place the nonreaching hand on the wheelchair for balance.

Special considerations for manual wheelchairs operated by caregivers (Version 2)
- None.

Special considerations for powered wheelchairs operated by users (Version 3)
- The power may be on or off. However, the spotter should intervene if he/she is concerned that the subject is about to move the wheelchair in a way that might result in the fingers being run over by the wheels.

Special considerations for powered wheelchairs operated by caregivers (Version 4)
- When bending or stooping to pick up the object, the caregiver may place the nonreaching hand on the wheelchair for balance.

Special considerations for scooters operated by users (Version 5)
- Scooter users usually get out of the scooter to pick up objects. This is safer than leaning from the seat, due to the high center of gravity and the possibility of a sideways tip.
- When bending or stooping to pick up the object, the scooter user may place the nonreaching hand on the wheelchair for balance.

Wheelchair Skills Training
General training tips
- See some of the general training tips for the "reaches high object" skill.
- The wheelchair user should use one hand on the wheelchair or thigh to help with balance and the other hand to pick up the object.
- For a wheelchair user with weak trunk muscles, to reach the ground he/she should move the arms to the thighs one at a time, and then to the feet, placing the chest on the thighs.
- Turning the object on its side may help to get a better grip.
- To make it easier to pick up the object, the wheelchair user may pull the object up against one of the wheels so that it does not move.
- If a wheelchair user has weak pinch strength, increasing the friction between the fingers and the object (e.g., by wearing gloves, wetting the fingers with saliva) can help to prevent dropping the object.

Special considerations for manual wheelchairs operated by users (Version 1)

- *Variations*
  - A moving pick-up can be accomplished if the wheelchair user holds the object against the bottom of the rear wheel with one hand as the wheelchair rolls forward, then both hands can be used to grasp the object when it rotates to the top of the wheel.

Special considerations for manual wheelchairs operated by caregivers (Version 2)

- To pick a dropped object off the ground, the caregiver may maneuver the wheelchair so that he/she can keep one hand on the wheelchair, for balance and control. Then, the caregiver can crouch and pick up the object with the other hand.

Special considerations for powered wheelchairs operated by users (Version 3)

- If the wheelchair can be repositioned (e.g., with respect to tilt, recline, seat height), this may be helpful.
- There is a danger of unintentionally rolling a wheel over the fingers or pinching the fingers between the drive wheel and fender. The safest approach is to first position the wheelchair, shut off the power, then pick up the object.

Special considerations for powered wheelchairs operated by caregivers (Version 4)

- As for Version 2.

Special considerations for scooters operated by users (Version 5)

- Scooter users most often stand and get out of the scooter to pick up objects. This is safer than leaning from the seat, due to the high center of gravity and the possibility of a sideways tip.
- When getting out of the scooter, the scooter user should keep at least one hand on the scooter for balance.

7.14 Relieves weight from buttocks

Versions applicable

Version	Wheelchair Type	Subject Type	Applicable
1	Manual	Wheelchair user	✓
2		Caregiver	✓
3	Powered	Wheelchair user	✓
4		Caregiver	✓
5	Scooter	Wheelchair user	

Skill level

- Basic.

Description

- The subject relieves weight from both buttocks, although not necessarily at the same time.

Rationale

- Weight relief is important for comfort and the prevention of pressure sores. Ideally, such relief should be performed often (e.g., every 20 minutes) and for prolonged periods of time (e.g., 2 minutes). However, for the purposes of assessment and training, a duration of a few seconds is considered representative of the subject's capability. This skill is not applicable for scooter users, most of whom can stand and sometimes walk short distances.

Prerequisites

- None.

Spotter considerations

- Spotter starting position: Near the wheelchair, on the side toward which the subject leans (if any).
- Risks requiring spotter intervention: Forward or sideway tip or fall when leaning.

Wheelchair Skills Test (WST)

Equipment

- None.

Starting positions

- Wheelchair user: In wheelchair, sitting upright.

Instructions to subject

- *"Take the weight off your bottom. Hold your position until I tell you to stop."*
- If a subject chooses to lean to one side, the tester may prompt the subject *"Now to the other side"* without penalty.

Capacity criteria
- As for the general scoring criteria, with the clarifications below.
- A "pass" should be awarded if:
  - Weight is relieved for 3 seconds.
  - While the weight is being relieved, the tester should be able to easily slide a hand between pressure-sensitive areas (the ischial tuberosities, coccyx, and greater trochanters) and the wheelchair or cushion. However, placing a hand into the pressure-sensitive areas is ordinarily not required for the WST and this should only be done with the permission of the subject. The tester must make his/her best judgment about the extent of the pressure relief achieved.
  - It is permissible for the wheelchair user to stand up, to bridge (lifting the buttocks by extending the legs, pushing the feet on the footrests or floor), to lean side to side (Figure 7.36), or to lean forward (Figures 7.37 and 7.38) to relieve pressure. If the wheelchair can be tilted or reclined to 30° or more, this is considered a pass, even though this is not as effective a means of pressure relief as leaning. The technique used should be recorded in the Comments section.
  - If the subject leans, he/she must lean forward or to both sides and needs to recover independently (e.g., using push-handles, armrests).
  - If the subject's wheelchair is fitted with an alternating pressure cushion, the tester needs to be convinced by palpation that there is adequate relief under the pressure points.
- A "pass with difficulty" should be awarded if the subject uses the "push-up" technique (Figure 7.39). The push-up method, applying forces to the armrests or seat to lift the buttocks straight up, requires more force

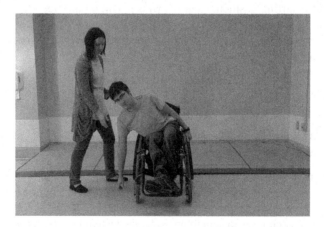

Figure 7.36 Using sideways leaning for the "relieves weight from buttocks" skill.

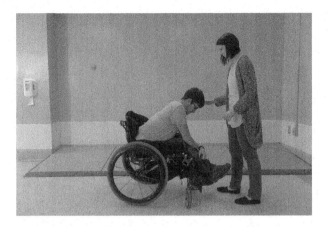

Figure 7.37 Leaning forward with elbows on knees to relieve weight from buttocks.

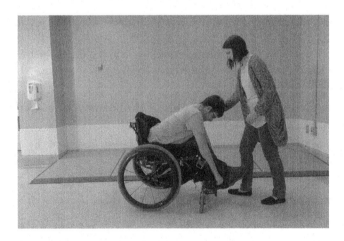

Figure 7.38 Leaning forward with chest on thighs to relieve weight from buttocks.

than some of the alternative methods. Over time, this may have adverse effects on the wheelchair user's wrists and shoulders. Also, this technique is difficult to maintain for the recommended length of time.

- A "testing error" score should be awarded if the tester is uncertain about the extent of weight relief and the subject refuses to permit the tester to perform a manual check.

Special considerations for manual wheelchairs operated by users (Version 1)

- It is permissible to do a tilt rest (see training section below) against a wall or other surface to meet the 30° tilt criterion.

139

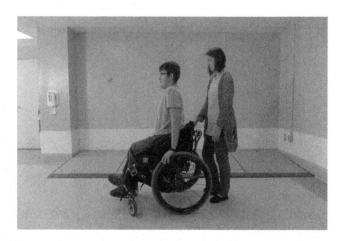

Figure 7.39 Using a "push-up" to relieve weight from buttocks (not recommended).

Special considerations for manual wheelchairs operated by caregivers (Version 2)
- The caregiver is only expected to assist (not completely perform) the wheelchair into and out of the weight-relieving position and to prevent any tips or falls.

Special considerations for powered wheelchairs operated by users (Version 3)
- None.

Special considerations for powered wheelchairs operated by caregivers (Version 4)
- As for Version 2.

Special considerations for scooters operated by users (Version 5)
- Not applicable.

Wheelchair Skills Training
General training tips
- It is generally suggested that a wheelchair user relieves pressure from the buttocks often (e.g., every 20 minutes) and for prolonged periods (e.g., 2 minutes). However, many wheelchair users have remained free of pressure ulcers with far less stringent regimens.
- If using any of the leaning methods, the extent of weight relief is proportional to the extent of the lean.
- With the forward-leaning method, the trunk can be rested on the thighs. Further unloading can be achieved by grabbing the footrests and pulling on them. Getting back upright from the forward-bent position can be a challenge for some wheelchair users. The hands can be walked up the thighs until an armrest or the backrest can be reached to allow the

person to pull him/herself the rest of the way. Leaning on a table is a strategy that may be helpful for wheelchair users who have difficulty in getting back to the upright position after leaning forward onto the thighs. It may be socially inconvenient to use the full forward-leaning technique in some circumstances. A more moderate forward lean, with the forearms resting on the thighs, may be adequate.

- Side leaning or shifting the weight onto one buttock can also be effective, for those who cannot lean forward and recover or in situations when the wheelchair user might find it inconvenient to lean forward. The armrests or rear wheels can be used to push or pull on. As was the case for leaning forward, the wheelchair user can lean on a table.
- After a weight-relief maneuver, the wheelchair user's buttocks should be gently repositioned on the seat rather than dropping them back into place.

- *Variations*
  - o Bridging, tilt, and recline are alternative methods that may be adequate for some wheelchair users. If tilt or recline are used, the greater the extent of tilt or recline the better. Anything less than 30° is probably inadequate.
  - o Standing up is effective, but if it is done using a stand-up wheelchair feature, there may be new pressure areas to consider related to how the wheelchair user is supported in the upright position. Standing on the footrests is generally not recommended, although it can be safe if the footrests are not too far forward and/or the casters are oriented in the forward-trailing position.
  - o Transferring out of the wheelchair (e.g., onto a bed), where the wheelchair user can lie on his/her side or front is also effective.
  - o As noted above in the WST criteria, push-ups are not recommended because of the high loads on the upper limbs (that may contribute to overuse symptoms) and because they cannot be sustained for long.

Special considerations for manual wheelchairs operated by users (Version 1)
- The leaning techniques can cause tips in the direction toward which the wheelchair user is leaning.
- If leaning forward fully, the casters should be in the forward-trailing position to increase forward stability.

- *Variations*
  - o A wheelie can be used to achieve tilt but the extent of tilt is usually much less than what is needed.
  - o The tilt-rest position (with wheel locks applied and the wheelchair leaning against a wall or curb) may permit sufficient rear tilt. This position can be achieved in different ways:

- Pull-back technique: The wheelchair user positions the wheelchair close to the object (e.g., a sofa, wall) that he/she intends to lean against. Some trial and error may be needed to select the correct distance from the object; it is better to start too close to the wall than too far from it. The wheel locks are applied and are checked to ensure that they are functioning. The wheelchair user then reaches back and pulls on the external object to tilt the wheelchair back just beyond the balance position, so that the wheelchair rests back against the object.
- Wheelie technique: The wheelchair user achieves the wheelie position with the back of the wheelchair facing the object that will be leaned against. The wheelchair is then rolled back in the wheelie position until the rear wheel or backrest of the wheelchair (for low and high objects, respectively) contacts the object. Then, the wheelchair is tilted back slightly further and the wheel locks are applied, one at a time. The wheelchair user must not let go of both wheels at the same time or the rear wheels will roll rapidly forward ("submarining") and a rear tip will occur.
- When returning from the tilt-rest to the upright position, the wheelchair user should leave the wheel locks on and tilt forward by leaning or by pushing against the object being leaned against.
  - Resting on the RADs may permit sufficient rear tilt but can result in a rear tip. With a spotter in place behind the wheelchair resting on the RADs, the wheelchair user can lean and rock backward to see whether the wheelchair tips over; if so, this technique should not be used.

Special considerations for manual wheelchairs operated by caregivers (Version 2)
- A caregiver can assist in a variety of ways, such as reminding the wheelchair user of the need to unload the buttocks or by assisting the wheelchair user in getting into or recovering from the unloaded position.
- A caregiver can sit behind the wheelchair and tilt the wheelchair backward to rest against the caregiver and provide pressure relief. To prevent the rear wheels from rolling forward, the wheel locks should be applied. This is a variation of the tilt-rest skill described above.

Special considerations for powered wheelchairs operated by users (Version 3)
- None.

Special considerations for powered wheelchairs operated by caregivers (Version 4)
- None.

Special considerations for scooters operated by users (Version 5)
- Not applicable.

7.15 Operates body positioning options

Versions applicable

Version	Wheelchair Type	Subject Type	Applicable
1	Manual	Wheelchair user	✓
2		Caregiver	✓
3	Powered	Wheelchair user	✓
4		Caregiver	✓
5	Scooter	Wheelchair user	✓

Skill level
- Basic.

Description
- The subject changes body position (i.e., tilts [Figure 7.40], reclines, elevates seat, elevates legrests, and/or uses the sit-to-stand feature) using the available options of a wheelchair and then restores the wheelchair to the original position.

Rationale
- Wheelchairs capable of variable body positions or postures are used for a variety of reasons, including pressure relief, to improve comfort, to ease breathing, to assist with postural control, to alter stability, to enhance transfers, to help overcome some obstacles, to facilitate bladder management, to reduce spasticity, or to reduce edema. Not all wheelchairs have body-positioning options.

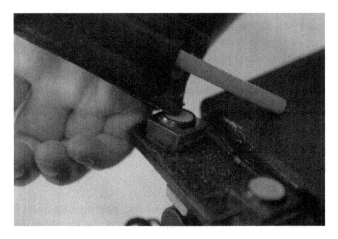

Figure 7.40 For the "operates body positioning options" skill, powered wheelchair user activating the tilt function.

143

Prerequisites
- None.

Spotter considerations
- Spotter starting position: Beside the wheelchair, in a position where it is possible to intervene. This position may vary depending upon the option being used.
- Risks requiring spotter intervention:
  - Runaway.
  - Damage to body parts or the external environment from the wheelchair mechanism.

Wheelchair Skills Test (WST)

Equipment
- None.

Starting positions
- Wheelchair: In whatever position the person is in so as not to demonstrate the skill while getting into a standard position.

Instructions to subject
- *"Show me how your wheelchair allows you to change body positions."*
- *"Bring the wheelchair back into the original position."*
- If there are other positioning options that have not been demonstrated, the tester may prompt the subject without penalty (e.g., *"Are there any other options that you can show me?"*).

Capacity criteria
- As for the general scoring criteria, with the clarifications below.
- A "pass" should be awarded if the subject successfully and safely changes body position in all of the ways possible for the wheelchair and returns to the original position.
- A "not possible" score can be awarded for this skill because not all wheelchairs have this capability.

Special considerations for manual wheelchairs operated by users (Version 1)
- None.

Special considerations for manual wheelchairs operated by caregivers (Version 2)
- None.

Special considerations for powered wheelchairs operated by users (Version 3)
- Note that some wheelchairs have apparent controls for which the wheelchair is not actually equipped.

144

Special considerations for powered wheelchairs operated by caregivers (Version 4)
- None.

Special considerations for scooters operated by users (Version 5)
- Some scooters allow the seat back to be reclined, slid forward and backward, and/or rotated to the side or back. If such options exist, the scooter user must be able to operate them to receive a "pass" score.

Wheelchair Skills Training
General training tips
- Before changing position or restoring the wheelchair to the original position, the person performing the action should check whether there is room behind the wheelchair and above the knees to change the position without damaging the environment, the wheelchair, the contents of a knapsack, the user, or a bystander.
- For wheelchair users with limited trunk balance, to reduce the likelihood of falling forward, 5°–10° of tilt or recline is usually adequate at rest or when driving.
- Depending upon the positioning mechanism, the extent of forward and rear stability of the wheelchair may differ in the new position. This should be taken into consideration when in a situation where reduced stability could be unsafe (e.g., proceeding forward up an incline in the tilted position) or when it might be helpful to alter the weight distribution between the front and rear wheels (e.g., to increase traction or reduce the tendency for smaller-diameter wheels to sink into a soft surface).
- If the wheelchair allows both tilt and recline, it is advisable to tilt first and then recline. When returning to the upright position, it is advisable to return from the recline position before recovering from the tilt position. This reduces the tendency for the wheelchair user to slide forward on the cushion.
- For wheelchairs that have stand-up and recline features, reclining the wheelchair user before standing him/her up may be preferable to standing up from the sitting position.

- *Progression*
  - For the wheelchair user to adjust to a position change may involve starting with a small position change and progressing to the full desired change.

Special considerations for manual wheelchairs operated by users (Version 1)
- None.

Special considerations for manual wheelchairs operated by caregivers (Version 2)
- When first tilting or reclining a chair, the caregiver should be aware of the force that may be required to "catch" the person at the desired angle. The setup and design of the tilt or recline mechanism influence the amount of weight supported by the caregiver.

Special considerations for powered wheelchairs operated by users (Version 3)

- *Adjustment tip*
  - Programming by the dealer and/or therapist should be considered to allow the wheelchair user to get into the desired position with as few steps as possible (e.g., using a preset position of 45° of tilt).
  - The wheelchair user needs to have access to the controller when in the altered position.
- For safety, some powered wheelchairs will prevent the wheelchair from being driven while in extreme positions. Powered wheelchairs may slow down or stop if the user attempts to operate them in unsafe circumstances (e.g., driving up a steep incline forward with the seat fully tilted back).
- Some seats can be turned to the side, allowing the powered wheelchair to be driven "sideways," such as along a table.
- Some seats can be turned completely backward, essentially converting a rear-wheel drive wheelchair into a front-wheel drive one and vice versa.
- When reversing the direction of the positioning option (e.g., from tilt back to tilt forward), it may be necessary to pause briefly with some controllers.

- *Progression*
  - If the rate of position change can be programmed, it is advisable to begin with a slow rate and progress to a faster one. This will provide more time in which to ensure that the wheelchair user is adjusting to the new position and that there are no body parts that are at risk of being injured.

Special considerations for powered wheelchairs operated by caregivers (Version 4)
- None.

Special considerations for scooters operated by users (Version 5)
- Some scooters allow the seat back to be reclined, slid forward and backward, and/or rotated to the side or back. If such options exist, they are usually carried out manually.

7.16 Level transfer

Versions applicable

Version	Wheelchair Type	Subject Type	Applicable
1	Manual	Wheelchair user	✓
2		Caregiver	✓
3	Powered	Wheelchair user	✓
4		Caregiver	✓
5	Scooter	Wheelchair user	✓

Skill level
- Basic.

Description
- The wheelchair user transfers from the wheelchair to another surface about the same height as the wheelchair seat and back again.

Rationale
- A transfer is a commonly used skill to move between the wheelchair and a chair, bed, tub, toilet, car, or other surface. The level wheelchair transfer should only be considered a representative transfer. More difficulty may be experienced when transferring to and from other surfaces or heights.

Prerequisites
- None.

Spotter considerations
- Spotter starting position: Usually in front of the wheelchair and slightly to one side, close enough to catch the subject if he/she falls and to prevent the wheelchair from rolling, sliding away, or tipping. The spotter may ask the subject where it would be best to stand, given the subject's previous experiences.
- Risks requiring spotter intervention:
  - Forward or sideways tip or fall when reaching or standing.
  - Rear tip when sitting back down in the wheelchair after a standing-pivot transfer.
  - Fall between the wheelchair and bench if the wheelchair rolls or slides away.
  - In the course of a standing-pivot transfer, tripping over the footrests and falling.

Wheelchair Skills Test (WST)

Equipment
- The following transfer surface is suggested (although any equivalent one is acceptable): a bench with a padded flat surface, no backrest and no

armrests. The sitting surface should be at least 1.0 m wide, 0.5 m deep, and 45–47 cm high. The bench legs should have rubber on their under-surfaces or other means to prevent the surface from moving.
- A transfer board (a piece of wood or plastic with beveled edges) should be made available for subjects who ordinarily use one. The subject may use his/her own equipment (if available).

Starting positions
- Wheelchair user: Seated in the wheelchair, and oriented in the chair as if he/she is ready to propel the chair (e.g., feet on footplates, if used).
- Wheelchair: Facing the bench and at least 0.5 m from it.
- The transfer board should be on the transfer bench within the subject's reach.

Instructions to subject
- Success on screening questions ("*Can you do it? How?*") is strongly rec-ommended before the subject is allowed to proceed to the objective test-ing of this skill.
- "*Transfer from the wheelchair to the bench* (indicate it)."
- "*Transfer back into the wheelchair.*"
- If, during the transfer, the subject is sitting on the target surface with the transfer board under him/her, it is permissible to prompt the subject to "*Move the transfer board away from you*" without penalty.
- After transferring back into the wheelchair, the subject may be prompted, without penalty, to restore the armrests or footrests.

Capacity criteria
- As for the general scoring criteria, with the clarifications below.
- A "pass" should be awarded if:
  o The wheelchair user is able to independently and safely set up the wheelchair for the transfer, transfer to and from the bench and restore the wheelchair to its operational condition.
  o Any safe and independent transfer technique is acceptable.
  o The transfer is not considered complete until the subject is off the transfer board.
  o The wheel locks, if any, may or may not be used.
  o If armrests need to be detached or moved out of the way for the transfer, after transferring back into the chair, they must be restored to their original positions.
  o If the wheelchair user's arm is secured to the arm support he/she must independently release and later replace his/her arm in the origi-nal position and state.
  o Although recommended, the subject need not clear the footrests if the transfer can be effectively and safely completed without doing so.

After transferring back into the chair, the footrests and feet should be as they were prior to the transfer.

o If a positioning belt is intended for independent use and is fastened around the wheelchair user at the beginning of the test, then the subject is expected to be able to undo it and fasten it again after transferring back into the wheelchair. If the wheelchair is equipped with a positioning belt, but the wheelchair user is not using it, the subject is not required to be able to use it.

o If the subject needs to reposition the unoccupied wheelchair between the transfer out of the wheelchair and the transfer back into it, the subject must do so him/herself.

• A "pass with difficulty" score should be awarded if:
    o During a standing pivot transfer, a 270° turn is used instead of using the shortest possible rotation.
    o The buttocks scrape significantly over the rear wheel or wheel-lock extension during a sideways transfer.
    o The lower limbs scrape over a footrest.
    o Poor ergonomic technique is used.

• A "fail" score should be awarded if:
    o The subject is the wheelchair user and the screening questions indicate that assistance is always required. There is no need to proceed to objective testing.
    o The subject falls onto the transfer bench and cannot get up without help.
    o The wheelchair user has a rear-closing seat belt or other restraint that is not intended for independent use, this is usually considered an automatic fail, unless the WST is being used to assess caregiver function.

Special considerations for manual wheelchairs operated by users (Version 1)
• None.

Special considerations for manual wheelchairs operated by caregivers (Version 2)
• The caregiver may receive physical assistance from the wheelchair user in performing the skill because it is not a reasonable expectation that a single caregiver could carry out this skill alone without additional equipment.

Special considerations for powered wheelchairs operated by users (Version 3)
• The controller may be on or off. Although the transfer is likely to be safer with the power off, the subject may need to move the wheelchair during the transfer and may not be able to control the power when not sitting in the wheelchair.

149

Special considerations for powered wheelchairs operated by caregivers (Version 4)
- As for Version 2.

Special considerations for scooters operated by users (Version 5)
- If the scooter seat can be swiveled to the side or back, this may be done.

Wheelchair Skills Training

General training tips
- There are a number of transfer techniques and surfaces to which a wheelchair user may wish to transfer. The methods described here are representative, but by no means comprehensive. Which type of transfer will be most suitable for a wheelchair user and/or caregiver will depend on a number of factors. An experienced clinician should make this determination. A thorough discussion of these options is beyond the scope of this book.
- Care should be taken to avoid catching the wheelchair user's catheter or other collection devices when transferring.
- The height of the starting and target surfaces should be adjusted, to the extent possible, such that the target surface is slightly lower.
- The path between the starting and finishing surfaces should be cleared of any obstacles.
- The wheelchair should be positioned as close as possible to the other surface, with the casters oriented in a way that enhances stability in the direction of transfer.
- The footrests should be cleared away (if possible) (Figure 7.41).
- The wheel locks should be applied.

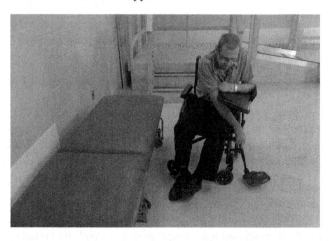

Figure 7.41 For the "level transfer" skill, swinging a footrest out of the way.

- The wheel locks of any other wheeled surface (e.g., bed) should be applied.
- The comments below generally apply to the transfer out of or into the wheelchair, but will be described as though the transfer is out of the wheelchair. Transfer into the wheelchair is generally the same except that, once the wheelchair user is back in the wheelchair, he/she should restore the footrests fully and put the feet back on them. The wheelchair user should also make sure that any removed or repositioned wheelchair parts (e.g., armrests, footrests, cushion, seat belt) are in the same position that they were before he/she left the wheelchair.

- *Sideways transfer*
    - This is sometimes called a "sliding" transfer but actual sliding is not recommended (to avoid shear forces or injury to the buttocks).
    - People using sideways transfers tend to lead with the weaker or more painful arm. However, if the arms are fairly symmetrical, alternating the leading and trailing arms allows them to share the stresses.
    - The wheelchair user should move the armrest (if any) out of the way on the bench side.
    - The wheelchair user should remove the wheel-lock extension (if any) on the bench side.
    - The feet should be supported on the floor if the footrests can be easily moved out of the way. If the footrests cannot be moved, it is acceptable to leave one or both feet on the footrests as long as forward tipping does not occur during the transfer. In addition to an actual tip, when the rear wheels become unloaded, the wheel locks become ineffective and the rear wheels may move sideways. This is less likely to cause a problem in wheelchairs that have the footrests behind the casters and will not happen if caster swivel locks (if any) are applied. In considering where to place the feet, the wheelchair user should try to avoid situations in which the feet are not free to swivel when the buttocks are moved to the new surface—this could lead to a torsion injury of the lower leg.
    - The wheelchair user should move forward on the seat, to avoid such obstacles to sideways movement as the rear wheels.
    - To get the transfer board (if using one) under the buttock, the wheelchair user should lean away from it.
    - The wheelchair user should push down on the transfer board and wheelchair to unload the buttocks.
    - The wheelchair user should avoid fully extending the fingers and wrists, allowing the fingers to wrap around the edge of the target sitting surface. This avoids overstretching the joints and tendons, that may be of importance for people with tetraplegia who use a tenodesis effect (whereby active wrist extension causes passive finger

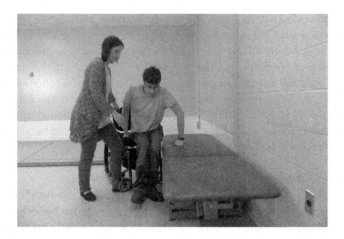

Figure 7.42 For the "level transfer" skill, midway through a sideways transfer.

flexion if the tendons are of appropriate length). Keeping the wrists in a neutral position also functionally lengthens the arms, making it easier to get the buttocks off the sitting surface.

o The wheelchair user should keep the leading hand just far enough away from the body to allow room for the buttocks to land on the target surface, but no farther. The trailing hand should be close to the body.

o The wheelchair user should shift sideways toward the target surface, in a single large movement (Figure 7.42) or several smaller ones.

o If possible, the wheelchair user should lean well forward ("nose over toes"). During the actual transfer from this position, the hips and the head move in opposite directions. For instance, if the wheelchair user wishes to move the buttocks up and to the right, the head should move down and to the left. This technique reduces the forces needed from the arms.

o Once the buttocks are fully supported by the target surface, the wheelchair user should remove the transfer board. The wheelchair user should lean away from it to do so.

o *Note:* It is not recommended that the wheelchair user place his/her feet on the bed or bench before independently attempting to move the buttocks sideways. Hamstring tightness will prevent the wheelchair user from being able to flex the hips adequately.

- *Standing-pivot transfer*
  o This is one of the most common types of transfer. The person stands fully upright from the original surface, pivots in place until his/her buttocks face the target surface, then sits down.

o Wheelchair users with hemiplegia using standing-pivot transfers tend to transfer to their stronger sides.
o The wheelchair user should leave the armrests in place.
o The wheelchair user should move forward on the seat before beginning the transfer.
o The wheelchair user should try to flex the knees to get the feet under the body, in preparation for the sit-to-stand phase of the transfer. During the sit-to-stand phase of the transfer, the hips should be flexed.
o To avoid the need for turning through a greater arc than necessary when pivoting, the wheelchair user should turn the back toward the bench rather than away from it.
o The wheelchair user may use the armrest to help maintain balance while transferring.
o If a wheelchair user with hemiplegia can only transfer back into the wheelchair with the strong side leading, he/she will need to move the wheelchair to the other side.

o *Variation*
  – The crouching transfer (Figure 7.43) is like the standing-pivot transfer, except that the knees and hips are not fully extended. The wheelchair user may need to move the armrest and the wheel-lock extension (if any) out of the way on the bench side. The wheelchair user should stay low, and not try to stand up fully. However, the buttocks need to be high enough to clear any obstacles (e.g., the armrest, rear wheel). The hips and the head move in opposite directions as for the sideways transfer.

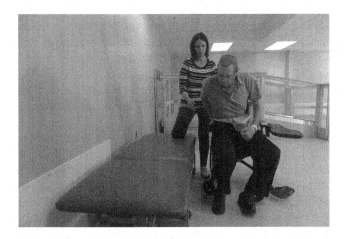

Figure 7.43 For the "level transfer" skill, midway through a crouching transfer.

- *Forward transfer*
  - When transferring straight-on (e.g., for a person with amputations of both legs above the knees), the wheelchair user should pull the wheelchair as close as possible to the transfer bench and at right angles to it.
  - A transfer board may be used.
  - The armrests should be left in place. If the armrests are desk-length, in some wheelchair designs, they may be reversed to provide better support as the wheelchair user moves from the wheelchair to the new surface.
  - Wheelchair users who have used the forward transfer method to transfer out of the wheelchair may be able to enter the wheelchair in the forward direction and then turn around, if their amputation residual limbs are short enough. Alternatively, they can back onto the wheelchair seat.

- *Progression*
  - Once the basic transfer is mastered, it should be practiced with different target surfaces, at different relative heights. The "gets from ground to wheelchair" skill discussed later is an extreme example of a transfer.

Special considerations for manual wheelchairs operated by users (Version 1)

- *Wheel locks*
  - Prior to the actual transfer, the learner should apply the wheel locks (if any). If the rear wheel is able to turn with the wheel lock applied, the wheel lock may need to be adjusted or the tire may need to be pumped up, if it is pneumatic. If strength is a limiting factor to applying the wheel locks, the wheelchair user may use wheel-lock extensions.
  - A wheelchair user with weak trunk muscles can avoid falling forward during wheel-lock handling, by hooking an arm around a push-handle or holding onto an armrest or wheel.
  - To apply a push-to-lock wheel lock, the wheelchair user grasps the handle of the wheel lock and pushes it toward the front of the wheelchair until firmly in place.
  - To apply a pull-to-lock wheel lock, the wheelchair user pulls the handle backward until firmly in place.
  - Retractable wheel locks are ones that can be positioned completely out of the way when they are not in use, so the wheelchair user does not scrape his/her hands on them during wheelchair propulsion. They are most often found on rigid-frame wheelchairs. To apply a retractable scissor wheel lock, the wheelchair user pulls or pushes the handle in the appropriate direction until firmly in place.

- o To release wheel locks, the subject should reverse the action used to apply them. For a retractable scissor wheel lock, the subject should fold the wheel lock fully out of the way.
- o For wheelchairs that are equipped with them, caster swivel locks can be used to help maintain caster orientation.
- *Armrests*
  - o Generally, it is easier to reposition the armrests than it is to remove them completely.
  - o To move the armrests away, any of the following options can be used, depending upon the armrest design:
    - For a flip-up armrest, the learner should unlock the front of the armrest from the receptacle and lift the front of the armrest so that it flips behind the chair back.
    - For a swing-away armrest, the learner should lift the armrest up slightly to disengage it and then swing it horizontally to the rear far enough to clear the backrest posts.
    - To completely remove an armrest, the learner should unlock whatever locks are necessary. There may be ones at both the front and back of the armrest. The learner should lift the armrest straight up so that the armrest is detached from the chair. If the armrest is height-adjustable, the wheelchair user should be careful not to just remove the elevating arm pad.
    - For a wheelchair with a tray (e.g., for a person with hemiplegia), the subject should first flip the tray away or slide it forward to detach it (Figure 7.44).

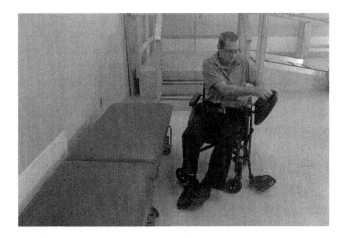

Figure 7.44 For the "level transfer" skill, flipping a half-tray out of the way.

- To restore the armrests: With some armrest designs, it is easy to unintentionally reverse left and right. To avoid this, the subject should be encouraged to follow a routine with respect to where the armrests are placed when removed, the subject should reverse the process used when moving the armrests away, the subject should make sure the armrest posts are lined up with the receptacles before locking them, and the subject should check to make sure the armrests are locked in place by pulling up on them.

- *Footrests*
  - The subject should clear the footrests out of the way prior to a transfer, whenever possible. It may be easier to do so before moving the wheelchair into its final position.
  - Before moving the footrests out of the way, the subject should first remove the feet from the footrests. A person with weak hands may need to use both hands or an extended wrist under the knee to lift the leg. If one leg is stronger, it may be used to assist in lifting the weaker leg. Later, after restoring the footrests, the subject should put the feet back on the footrests.
  - To move a swing-away footrest out of the way, the subject should unlock the footrest. Locking mechanisms vary from wheelchair to wheelchair. The subject should swing the footrest completely out of the way. Some footrests swing away to the side and others to the middle. To replace the footrest, the subject should push the footrest back toward the front of the wheelchair until it clicks into place. The subject should check that it is locked in place by pulling on it.
  - To completely remove the footrests, the subject may need to first swing the footrest away. The subject should then pull up on the footrest. The subject should pay attention to how the footrest was attached to the chair to simplify restoring it later. To replace the footrest, the subject may need to start in the swung-out position, line up the post or pins with the hole(s), and put the footrest back in place. The subject should then swing the footrest back to the front.
  - Some wheelchairs do not allow the footrests to be swung away or removed, but it may be possible to flip the foot-plates up. The subject should pull the foot-plates up until they are fully vertical. To do so on some wheelchairs, it may be necessary to push the heel loops (if any) forward. To replace the footrests, the subject should push the foot-plates down. The learner should push the heel loops back into place, if they were displaced earlier.
  - To raise an elevating footrest, the subject should grasp it near the end and lift it to the desired position. This requires less force if the leg

is not on the leg-rest. To lower the footrest, the subject should support its weight, and hold the position lock open while lowering the footrest. The position lock is often located at the top of the leg-rest (near the knee).

o For a wheelchair user with weak trunk muscles, to reach the footrests, the arms can be moved to the thighs one at a time, and then to the feet, until the chest is resting on the thighs. To get back into the upright position, the stronger arm can be hooked over the push-handle or armrest and the body pulled up.

• If possible, the subject should position the wheelchair so that the casters are trailing in the direction of the transfer to reduce the likelihood of the wheelchair tipping in that direction. To achieve this position, the subject should finish the wheelchair positioning with a slight movement away from the direction of the transfer.

Special considerations for manual wheelchairs operated by caregivers (Version 2)

• This section only deals with transfers for wheelchair users who require minimal assistance to perform the final movement between the wheelchair and the bench. If the caregiver must perform the majority of the effort, or if a mechanical lift is needed, additional training by experienced rehabilitation professionals is needed. This is outside the scope of this book.

• The caregiver should be attentive to the position of the wheelchair user's arms to avoid injuring them during the transfer.

• The caregiver should inquire as to whether the wheelchair user has ever experienced falls and, if so, in which direction. This may help the caregiver to know how best to provide assistance.

• Care should be paid to good back ergonomics for the caregiver:

o The feet should be shoulder-width apart for balance.

o The caregiver should avoid bending the back and twisting at the same time.

o The caregiver should bend his/her knees and keep the rest of his/her body straight to avoid injury to the back.

o The caregiver should keep the wheelchair user close to the caregiver (vs. arms straight).

o The caregiver should involve the wheelchair user as much as possible.

o The caregiver should use aids (e.g., transfer belt, transfer board, mechanical lift) as needed.

o The caregiver should use the help of other people, if help is needed. One option is for one caregiver to be behind the wheelchair user, reaching under the upper arms to grasp the wheelchair user's forearms that have been crossed in front of the body. The second caregiver is positioned in front of or to the side of the wheelchair and lifts the legs from behind the knees.

o The wheelchair user should not hold the caregiver around the neck.

o If the wheelchair user is falling, it may be necessary for the caregiver to lower him/her to the floor rather than risk injury to the caregiver.

o If it is necessary for the caregiver to move the unoccupied wheelchair to the other side, the caregiver may leave the wheel locks on. Using the push-handles at the rear of the wheelchair, the caregiver should lift the rear wheels slightly off the floor and push or pull the wheelchair on the casters (the "wheelbarrow" method). This will save time, avoid strain on the back, and ensure that the wheel locks are applied when the wheelchair user transfers back into the wheelchair. Because the only wheels on the floor are the casters, the wheelchair can be moved straight sideways.

- *Sideways transfers*
  o It may be necessary to perform the transfer in steps.

- *Standing-pivot transfers*
  o To assist the wheelchair user in getting from sitting to standing, the caregiver should stand or sit in front of the wheelchair or stand to one side.

  o The caregiver should apply an assisting force to the wheelchair user's body, near the hips. The caregiver should not pull on the wheelchair user's arms.

  o The caregiver may use a transfer belt around the wheelchair user's waist.

  o The caregiver may need to use his/her knees to keep the wheelchair user's knees from buckling, by blocking them.

  o Once standing, the caregiver should ask the wheelchair user to pivot, turning the back, in the shortest possible route, toward the bench.

Special considerations for powered wheelchairs operated by users (Version 3)

- Positioning (e.g., tilt, recline, seat height, seat swivel) may be useful while preparing the wheelchair for the transfer.
- The power should generally be turned off while the transfer is being performed.
- Although not the only consideration, if all other factors are equal, it will be easier to make a sideways transfer toward the non-joystick side.

- The controller may need to be moved out of the way for a sideways transfer.
- If the wheelchair user is using a standing pivot transfer, with the feet on the ground, the tilt mechanism of the wheelchair can be used to assist in lifting the buttocks if the wheelchair user has moved well forward on the seat.

Special considerations for powered wheelchairs operated by caregivers (Version 4)

- If a mechanical lift is being used, it can be helpful to put the seat in the tilted position to assist in ensuring that the wheelchair user is properly positioned in the sling.
- If a mechanical lift is being used, after the wheelchair user has been lifted sufficiently, it may be easier to drive the wheelchair backward out from under the wheelchair user rather than to move the lift.

Special considerations for scooters operated by users (Version 5)

- The tiller handles can be an asset while transferring if the scooter user needs assistance for balance. However, the amount of force applied to them should be minimal because they may swivel into a different position.
- The handles can get in the way if the tiller is turned to the side toward which the scooter user is transferring (Figure 7.45).

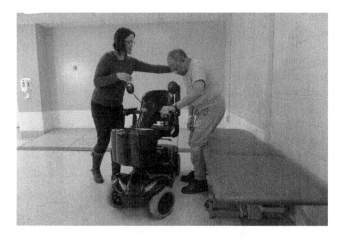

Figure 7.45 For the "level transfer" skill, midway through a standing-pivot transfer with the scooter tiller turned away.

7.17 Folds and unfolds wheelchair

Versions applicable

Version	Wheelchair Type	Subject Type	Applicable
1	Manual	Wheelchair user	✓
2		Caregiver	✓
3	Powered	Wheelchair user	
4		Caregiver	
5	Scooter	Wheelchair user	

Skill level
- Intermediate.

Description
- The subject folds or takes apart the unoccupied wheelchair to make it as small as possible, and then restores it to its original condition.

Rationale
- For transport or storage, the size or weight of the wheelchair may need to be reduced. This can be done by folding the wheelchair. Removal of the rear wheels or other parts is a useful way to further diminish the size and weight of the wheelchair. For the purposes of the WSP, this skill is considered "not applicable" for powered wheelchairs and scooters, even though some parts of some models may be readily foldable or removable.

Prerequisites
- None.

Spotter considerations
- Spotter starting position: Near the subject, on the side toward which the subject leans (if any).
- Risks requiring spotter intervention:
  - Forward tip or fall while reaching.
  - Pinching fingers between folding or rotating parts.
  - Injury to lower leg or foot due to dropping wheelchair parts.

Wheelchair Skills Test (WST)

Equipment
- Surface for the subject to sit on, if needed.

Starting positions
- Wheelchair user: Seated or standing beside the wheelchair.

- Wheelchair: In the same position and condition as immediately after the wheelchair user had transferred out of it. If the subject has removed some wheelchair parts (e.g., an armrest, footrest) as part of the transfer out of the wheelchair and restores the wheelchair to its original state after transferring back into it, the tester may consider these actions as part of the "folds and unfolds wheelchair" skill.

Instructions to subject

- *Note*: This skill is usually assessed with the "level transfer" skill, while the wheelchair user is out of the wheelchair.
- *"Fold the wheelchair as tightly as you can or take it apart, as if you were going to store it."*
- *"Put the wheelchair back together and open it so that you can get back into it."*

Capacity criteria
- As for the general scoring criteria, with the clarifications below.
- A "pass" should be awarded if:
  o The wheelchair is folded or taken apart such as to reduce the dimensions of the wheelchair as much as possible without tools. If the wheelchair is incompletely folded or taken apart, it is acceptable to prompt the subject without penalty (e.g., *"Can you get it a little tighter or smaller?"*, *"What if it was still too big or heavy?"*).
  o If wheelchair components or accessories (e.g., cushion, rigid seat, backrest, knapsack, footrests) need to be removed to achieve the smallest dimension, this should be done.
  o For a rigid wheelchair with a backrest that folds forward, the backrest canes and the seat rails should be as close to parallel with each other as is mechanically possible. If the cushion prevents this, the tester may prompt the subject by asking *"Can you get this folded more tightly?,"* but the tester must not suggest the solution of removing the cushion.
  o The rear wheels should be removed if this can be done without tools (i.e., if they are of the quick-release type). If the rear wheels are not removed spontaneously by the subject, he/she may be prompted to do so without penalty (e.g., *"Do the rear wheels come off? Can you show me?"*).
  o It is acceptable for the subject to use the foot to help fold and unfold the wheelchair.
  o For the unfold component of the skill, the wheelchair should be opened fully.

    o If the wheelchair is incompletely restored to its original condition, it is permissible, without penalty, to cue the subject by inquiring *"Is the wheelchair in the same condition that it was in before you folded it?"*

    o After putting the rear wheels back on the frame, the subject should check that they are firmly in place by pulling on them.

- A "pass with difficulty" should be awarded if the test subject puts a contoured cushion in backward, because of the potential for causing a pressure sore.
- A "fail" score should be awarded if:
  - The subject does not know that the wheelchair folds or that the rear wheels are removable without tools.
  - The wheelchair has been opened in a way that precludes full use of the wheelchair (e.g., by tangling a seat-belt strap in a way that will cause it to rub on a wheel, seat rails not sitting in rail saddles). The tester should correct the problem before the wheelchair user gets back into the wheelchair.
- A "not possible" score can be awarded for this skill because not all wheelchairs have this capability.
- A "testing error" score can be awarded for this skill if the "level transfer" cannot be achieved independently and the tester is unable to assist the wheelchair user out of the wheelchair.

Special considerations for manual wheelchairs operated by users (Version 1)
- For a wheelchair with a spotter strap around a cross-brace, the strap may be removed by the tester, without penalty, to permit the wheelchair to fold fully.

Special considerations for manual wheelchairs operated by caregivers (Version 2)
- None.

Special considerations for powered wheelchairs operated by users (Version 3)
- Not applicable.

Special considerations for powered wheelchairs operated by caregivers (Version 4)
- Not applicable.

Special considerations for scooters operated by users (Version 5)
- Not applicable.

Wheelchair Skills Training
General training tips

- *Fold wheelchair*
  - The subject should pay attention to each item as he/she removes or alters it, to ensure that he/she will be able to reassemble the chair later.

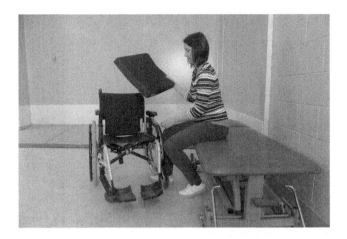

Figure 7.46 For the "folds and unfolds wheelchair" skill, removing the seat cushion.

- o The subject should remove anything that may prevent folding (such as the cushion, rigid seat, backrest, knapsack) (Figure 7.46).
- o To remove a rigid seat or backrest, the subject may need to release restraining devices.
- o For rear wheels that can be removed without tools (Figure 7.47), there is usually a release mechanism at the center of the axle, a button or lever that needs to be depressed. If the wheel does not come off easily, the subject should check to make sure the wheel lock is not on and that the rear wheel is off the ground.

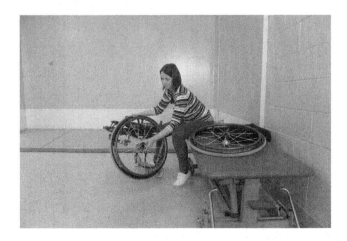

Figure 7.47 Removing the rear wheels.

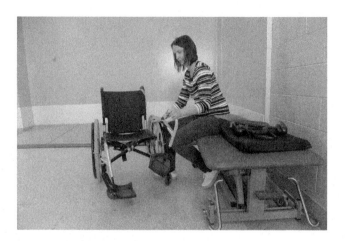

Figure 7.48 Removing a footrest.

- To fold a cross-braced wheelchair (one that becomes narrower from side to side when folded), the subject should first clear the footrests (e.g., by flipping them up, swinging them away, removing them) (Figure 7.48).
- To fold a cross-braced wheelchair more easily, the subject should position the wheelchair so that he/she is on one side of it. The subject should then tip the chair slightly toward him/herself so that the wheels on the side away from him/her are off the ground (Figure 7.49). This eliminates the friction of the far-side rear wheels on the

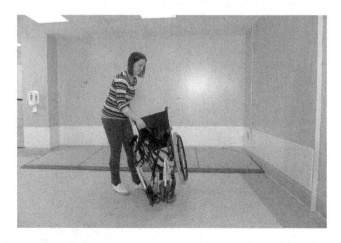

Figure 7.49 For a cross-brace folding wheelchair, tilting the chair to one side and pulling up on the sling seat to fold the chair.

164

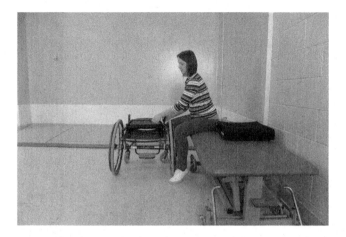

Figure 7.50 Folding down the backrest of a rigid-frame wheelchair.

ground and allows gravity to assist in folding the wheelchair. The subject should then pull the seat or seat rails upward, with one or both hands, to fold the chair.

o For a rigid-frame wheelchair with a fold-down back, although the frame cannot be folded, the subject can often make the chair easier to transport by folding down the back (Figure 7.50). The learner may need to release any restraining devices before he/she can do so.

o The push-handles of some wheelchairs can be folded to further reduce the wheelchair dimensions.

o After folding the wheelchair, if the wheelchair does not have a latch mechanism to prevent the wheelchair from opening while it is being lifted, it may be helpful to use a strap.

o When lifting a folded wheelchair for which the rear wheels cannot be removed, injury can occur if the unlocked rear wheels are grasped, because the frame will be free to rotate.

- *Unfold wheelchair*
  o Generally, the subject should reverse the steps used to fold the wheelchair and in roughly reverse order (e.g., starting by putting the rear wheels back on and finishing with putting the cushion back in place).

  o If the rear wheels have been removed, they should be replaced. It may be necessary to push the quick-release plunger to allow the axle to get into the housing. To check that the axle is fully seated, the plunger should be out and it should not be possible to pull the rear wheel off.

  o Some tires have a directional tread pattern (more rolling resistance in one direction than the other). If so, the left and right wheels should not be considered interchangeable.

Figure 7.51 When opening a cross-brace folding wheelchair, pushing down on the seat rails with open hands.

- o The subject should be careful not to tangle the seat belt (if any) under the seat.
- o To get the process of opening a cross-braced wheelchair started, the subject can lift the rear wheels off the ground and separate the push-handles.
- o The subject usually needs to push the seat rails back down into the starting position. The subject should keep the fingers on top of the rails to prevent them from being pinched (Figure 7.51).
- o For wheelchairs with backrests that fold forward, the backrest may lock in the folded position, necessitating a release of the locking mechanism to unfold the backrest.
- o The subject should remember to put the cushion back on the seat properly before transferring back into the chair.

- *Progression*
  - o Once the learner is able to fold and unfold the wheelchair, he/she can progress toward full use of this skill by putting the folded wheelchair up on the transfer bench and into his/her vehicle. Variations in the designs of the wheelchair and vehicle preclude a thorough discussion of this in this book.

Special considerations for manual wheelchairs operated by users (Version 1)

- *Variations*
  - o The advanced wheelchair user may be able to remove and replace rear wheels while seated in the wheelchair by leaning sideways (e.g., in a doorway) or forward (tipping the wheelchair onto the footrests).

Special considerations for manual wheelchairs operated by caregivers (Version 2)
- None.

Special considerations for powered wheelchairs operated by users (Version 3)
- Although generally not applicable, some powered wheelchairs can be folded or reconfigured without tools for storage or transportation. If that is the case and doing so is a goal of the subject, training should be provided.

Special considerations for powered wheelchairs operated by caregivers (Version 4)
- As for Version 3.

Special considerations for scooters operated by users (Version 5)
- As for Version 3.

7.18 Gets through hinged door

Versions applicable

Version	Wheelchair Type	Subject Type	Applicable
1	Manual	Wheelchair user	✓
2		Caregiver	✓
3	Powered	Wheelchair user	✓
4		Caregiver	✓
5	Scooter	Wheelchair user	✓

Skill level
- Intermediate.

Description
- The subject opens, passes through, and closes a hinged door that opens away from the subject, then repeats the task in the opposite direction (with the door opening toward the subject).

Rationale
- Wheelchair users frequently encounter such hinged doors or gates. Although there are a variety of door types, this is considered a representative skill.

Prerequisites
- None.

Spotter considerations
- Spotter starting position: Near the wheelchair, on the side toward which the subject leans (if any).
- Risks requiring spotter intervention:
  o Forward or sideways tip or fall due to reaching and pulling on the door handle.
  o Pinching the fingers between the door and the frame.

Wheelchair Skills Test (WST)

Equipment
- Door approximately 81 cm wide, preferably with no resistance to opening.
- Preferably a lever handle greater than 10 cm in length and 75–90 cm above the floor.
- Preferably no threshold (because this is evaluated separately later).
- There should be enough space (preferably at least 1.5 m square), on both sides of the door, to allow the subject to maneuver.

Starting positions
- Wheelchair: Facing the closed door with the front wheels at least 0.5 m from it.

Instructions to subject
- *"Open the door, move the wheelchair through it and close the door behind you."*
- *"Now, go back through the door the other way."*
- The order of performing the two components of this skill test is not important.
- If the subject leaves the door slightly ajar, he/she may be prompted, without penalty, to finish closing it.

Capacity criteria
- As for the general scoring criteria, with the clarifications below.
- A "pass" should be awarded if:
  - The skill in each direction is completed (when the door closes firmly).
  - Any safe and effective technique is used. For instance, the subject may close the door by reaching back for it. Alternatively, the subject may proceed away from the door and then turn around and come back to close it.
- A "fail" score should be awarded if a finger pinch seems likely between the door and the frame. The spotter should intervene to prevent injury.

Special considerations for manual wheelchairs operated by users (Version 1)
- The subject may use the door frame to assist in passing through the door.

Special considerations for manual wheelchairs operated by caregivers (Version 2)
- None.

Special considerations for powered wheelchairs operated by users (Version 3)
- None.

Special considerations for powered wheelchairs operated by caregivers (Version 4)
- None.

Special considerations for scooters operated by users (Version 5)
- None.

Wheelchair Skills Training

General training tips

- *Adjustment tips*
  - For doors in the wheelchair user's own environment, attaching something (e.g., a handle, piece of rope) in the middle of the door can make closing easier.
  - Having a roller on the outer corner of the wheelchair's footrest can be useful if using the footrests to apply force to a door.
- Although the footrests can be useful to help push doors open or closed, this method should not be used on glass doors that might break.

- If using the footrests to apply a force to a door, it is best to approach the door at a slight angle toward the side that will open. This ensures that it is the outer corner of the footrest that contacts the door and not the feet.
- The feet often extend out beyond the footplates, so care needs to be taken to avoid injury.
- If there is a threshold in the doorway, the principles for dealing with such an obstacle can be found in the later section on the "gets over threshold" skill.
- For a door that opens away from the wheelchair, the wheelchair user can begin the skill by positioning the wheelchair directly in front of the door (Figure 7.52).
- For a door that opens toward the wheelchair, the wheelchair user should position the wheelchair to the side of the door to allow room for it to be swung open without striking the wheelchair or a body part.
- Once a self-closing door has been opened enough to allow the wheelchair to proceed through it, the widest part of the wheelchair can be used to prevent the door from closing. To avoid scraping the door, the wheelchair user can use his/her hand or elbow to push the door open briefly to allow progress.
- While moving past the door, the wheelchair user should be careful to avoid catching any clothing or body parts on the door handle or the surface of the door if it is rough.
- To close a door that opens toward the wheelchair, after passing through it, there are several options (if the door does not close by itself):
  o The wheelchair user may gently swing the door closed behind him/her, moving the wheelchair quickly through the door and out of the way (Figure 7.53a through c).

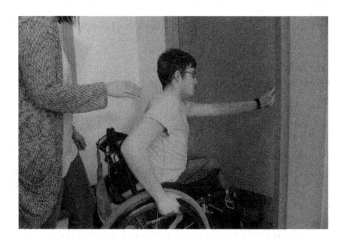

Figure 7.52 For the "gets through hinged door" skill, positioning for a door that opens away.

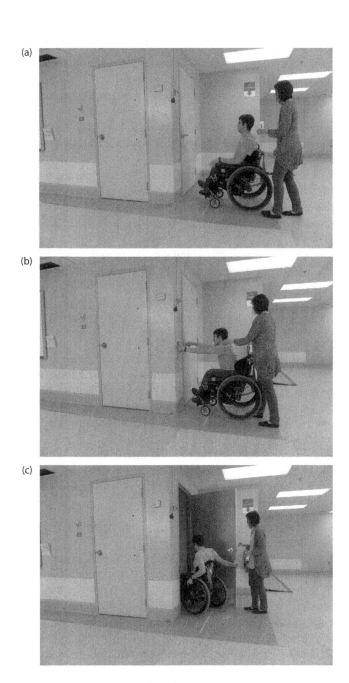

Figure 7.53 For the "gets through hinged door" skill for a door that opens toward the wheelchair user: (a) starting position, (b) left hand on door frame, and (c) pulling the door closed as he passes through.

- o The wheelchair user may turn around once through the doorway, reach forward and pull the door toward him/her while backing away (Figure 7.54a through c).
- o The wheelchair user may go through the door backward, pulling the door with him/her.
- o The wheelchair user should not put his/her fingers between the door and door frame for any longer than necessary because they may get pinched when the door closes.
- o Reaching over the back of the wheelchair to close the door is effective, but there is risk of a rear tip in a manual wheelchair.
- To close a door that opens away from the wheelchair after passing through it, there are several options (if the door does not close by itself):
  - o The wheelchair user can swing the door closed.
  - o The wheelchair user can turn the wheelchair around and push the door closed with the footrests.
  - o The wheelchair user can back up to close the door using the rear wheel or other wheelchair part to push on the door.

- *Progression*
  - o Judging the width of doorways relative to wheelchair dimensions can require practice. To avoid damage to the hands, wheelchair, or door frame, it can be useful to attempt getting through progressively more narrow openings using objects that are not firmly fixed (e.g., pylons). Bubble wrap can be used to provide audible feedback.
  - o The subject should start with a door that does not close on its own and progress to one that does. The trainer can reduce or add resistance to door opening by applying forces with his/her hand.

- *Variations*
  - o The learner can experiment with negotiating the door in the forward or backward directions.
  - o Game: There are many variations in the ways doors open and close, alone or in sequence with other doors. Also a variety of door handles exist. A game that provides opportunities to practice these variations is to have a "door scavenger hunt," seeing how many different combinations and permutations can be found and successfully managed in a period of time.
  - o Game: To get used to the relative widths of the wheelchair and doorways, the user can attempt to get between two obstacles that are lightweight and movable enough that injury is not a concern. The distance between the obstacles can be varied and they can be approached at progressively greater speeds.

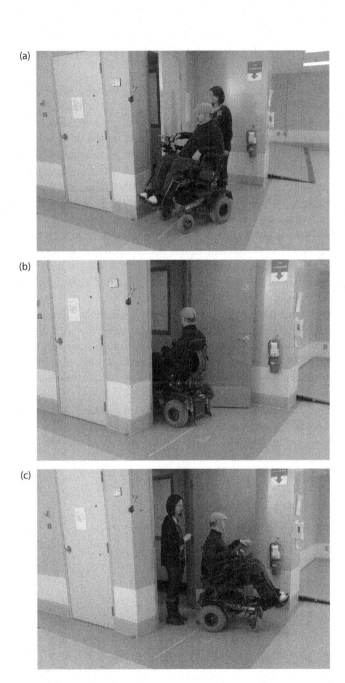

Figure 7.54 For the "gets through hinged door" skill for a door that opens toward the wheelchair user: (a) positioning the wheelchair with room for the door to swing open, (b) passing through the opening, and (c) coming back to close the door.

173

Special considerations for manual wheelchairs operated by users (Version 1)

- *Two-hand-propulsion pattern*
  - The door frame can be used to help propel the wheelchair user through the door (the "slingshot" method) (Figure 7.55). To do so, the wheelchair user reaches forward and places one hand on the door frame and the other on the door or the door frame on the other side. Then, by pulling with both hands, the wheelchair is moved through the opening. This has the advantage of keeping the hands from being injured by bumping or scraping them between the door frame and the wheelchair.
  - To open a door that opens away from the wheelchair more easily, the wheelchair user can turn sideways in front of it. This allows the wheelchair user to get closer to the door and to resist the tendency of the wheelchair to roll backward when the door is pushed. Alternatively, the wheelchair user can hold onto the door frame with one hand, as the door is pushed with the other. This is more likely to be necessary if the door resists opening.
  - To open a door that opens toward the wheelchair more easily, the wheelchair user should push on the door frame with the hand farthest from the hinge to open the door more easily with the other hand. Turning the wheelchair sideways will also prevent the wheelchair being pulled forward as the wheelchair user pulls on the door.
  - The wheelchair user may keep one hand on the door handle and use the other hand to push both wheels, one at a time. This is slow and awkward but may be effective for some wheelchair users.

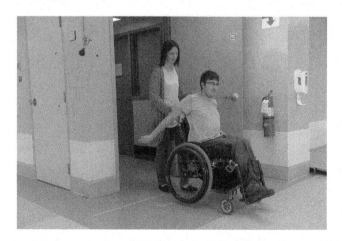

Figure 7.55 Using the door frame to propel the wheelchair through the opening ("slingshot" method).

- *Variations*
  - o If there is a threshold or level change in the door opening, it may be helpful to use the door frame to help provide the forces needed to proceed.
  - o For a doorway that is too narrow for the wheelchair to pass through, an option is for the wheelchair user to transfer from the wheelchair on one side of the door to a regular chair on the other, fold the wheelchair to get it through the door and then transfer back into it. Other alternatives include removing both rear wheels and resting on the RADs or transport wheels to get through the door. For wheelchairs that fold from side to side, some wheelchair users can partially fold the wheelchair and sit on an armrest.
  - o For a door that opens away from the wheelchair and that is latched with a bar mechanism that will open when a force is applied to it (Figure 7.56), the wheelchair user can approach the door without slowing down. At the last moment, the wheelchair user can lean and reach forward with one or both hands and use momentum to open the door. The feet should not strike the door. This should be practiced at slow speeds initially.

- *Hemiplegic-propulsion pattern*
  - o Using one hand to cross over from one wheel to the other can be helpful to keep the wheelchair straight while getting through a door.

Figure 7.56 Using a moving approach, applying force to a push bar of a door that opens away from the wheelchair user.

Special considerations for manual wheelchairs operated by caregivers (Version 2)

- Before pushing a wheelchair through any type of door or narrow space, the caregiver should make sure that the wheelchair user's hands or elbows are not extending beyond the sides of the wheelchair where they could be injured.
- The caregiver should keep part of his/her body between the door and the wheelchair user.
- The skill can be accomplished by moving the wheelchair through the door forward or backward.
- For a narrow doorway, one option is for the caregiver to remove one rear wheel. With the wheelchair user leaning the other way and the caregiver supporting the push-handle, it may be possible to get through the door on three wheels.
- For a door that opens away from the wheelchair, the caregiver should open the door, grasp the push-handles at the rear of the wheelchair, and push or pull the wheelchair through the doorway. When the wheelchair and caregiver are completely out of the way, the caregiver should close the door.
- For a door that opens toward the wheelchair, if there is room, the caregiver should angle the wheelchair away from the door on the side that will open.

Special considerations for powered wheelchairs operated by users (Version 3)

- When applying a force to open a door toward a powered wheelchair, it may be easier to simply grasp the door handle with the hand on the side away from the joystick and then back the wheelchair up, rather than doing all of the work with the arm.
- Unlike with a manual wheelchair, the force of a self-closing door does not require the user to brace himself/herself with the other hand on the door frame or to turn sideways to prevent the wheelchair from being moved unintentionally.
- For a person with hemiplegia, it is impossible for the sound arm to simultaneously hold the door lever and control the joystick. It may be necessary to complete the task in several small steps.
- Because of the risk of injury and because overcoming the force of a self-closing door mechanism is not a problem, it is not recommended that momentum be used to open doors with latch mechanisms.
- If the powered wheelchair is about to collide with the door or door frame, the wheelchair user should not reach out to fend off with the hands or feet—this is ineffective and may cause injury. The body parts should generally be kept within the protective envelope of the wheelchair.

Special considerations for powered wheelchairs operated by caregivers (Version 4)

- This can be an awkward task, because the caregiver's position is dictated by both the need to have access to the joystick and the door.

Special considerations for scooters operated by users (Version 5)

- The width of some scooters may make it difficult to get them through narrow openings.
- The length of some scooters can make it difficult to reach door handles (Figure 7.57), making it necessary for the scooter user to get off the scooter. When getting off the scooter, the user should keep at least one hand on the scooter for balance. The scooter user should not operate the scooter while standing because the movement may cause a fall.

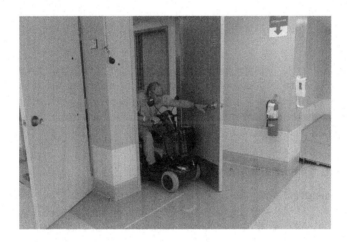

Figure 7.57 Scooter user having difficulty reaching the door handle to pull the door closed.

7.19 Rolls longer distance

Versions applicable

Version	Wheelchair Type	Subject Type	Applicable
1	Manual	Wheelchair user	✓
2		Caregiver	✓
3	Powered	Wheelchair user	✓
4		Caregiver	✓
5	Scooter	Wheelchair user	✓

Skill level
- Intermediate

Description
- The subject moves the wheelchair a longer distance on a smooth level surface. This may be done in the forward or backward direction.

Rationale
- The ability to manage longer distances allows wheelchair users to get around in the community (e.g., getting from a parking lot to an office, getting around inside a store). Subjects who are able to move their wheelchairs short distances may not be able to roll longer distances due to the additional endurance or attention required.

Prerequisites
- "Rolls forward short distance" skill.

Spotter considerations
- Spotter: If the subject has already safely performed the appropriate shorter-distance skill (forward or backward), the spotter only needs to be nearby.
- Risks requiring spotter intervention:
  o As for the appropriate shorter-distance skill (forward or backward).
  o Because speeds are usually faster if the skill is being performed in a smooth open space, the higher momentum can cause greater injury or damage if there is a collision with a fixed or moving obstacle.

Wheelchair Skills Test (WST)

Equipment
- A smooth level surface at least 1.5 m wide and 50–100 m long is ideal. Using multiple laps of a shorter distance is permissible, but it is preferable for the straight stretches to be at least 25 m in length, to minimize the number of turns. A curved path may be used.
- Space at least 1.5 m before the starting line and beyond the finishing line.

Starting positions
- Wheelchair: Leading wheel axles facing and behind the starting line.

Instructions to subject
- *"Move the wheelchair to the finish line* (indicate it or the number of laps)."
- If the subject moves outside the boundaries of the corridor to avoid a collision (e.g., during the "avoids moving obstacles" skill that is usually tested in conjunction with the "rolls longer distance" skill), he/she may be prompted, without penalty, to return to the corridor boundaries and to continue with the "rolls longer distance" skill.

Capacity criteria
- As for the general scoring criteria, with the clarifications below.
- Generally, as for the "rolls forward short distance" skill except:
- A "pass" score may be awarded if the subject propels the wheelchair backward.
- A "pass with difficulty" should be awarded if the subject develops symptoms of overexertion.
- A "fail" score should be awarded if the subject has failed the "rolls forward short distance" skill if a forward propeller or if the subject has failed the "rolls backward short distance" skill if a backward propeller.

Special considerations for manual wheelchairs operated by users (Version 1)
- None.

Special considerations for manual wheelchairs operated by caregivers (Version 2)
- None.

Special considerations for powered wheelchairs operated by users (Version 3)
- None.

Special considerations for powered wheelchairs operated by caregivers (Version 4)
- None.

Special considerations for scooters operated by users (Version 5)
- None.

Wheelchair Skills Training
General training tips
- As for the "rolls forward short distance" or "rolls backward short distance" skill, depending upon the direction used.
- The wheelchair user should keep the wheelchair away from dangers like walls or drop-offs, especially if the subject is one who regularly drifts to one side.

- *Progression*
  - ○ Start at a slow speed and increase as tolerated.
  - ○ Start in a smooth level indoor space and progress to the outdoor setting.

- *Variations*
  - ○ To work on directional control, the learner can follow a wall or side-walk edge while trying to stay within an arm's reach.

Special considerations for manual wheelchairs operated by users (Version 1)

- *Adjustment tip*
  - ○ If tire pressure is low, more effort will be needed.
  - ○ Solid tires roll better on smooth surfaces but are less comfortable than pneumatic tires on rough ground.
- Endurance may be a limiting factor if the wheelchair user is poorly conditioned.
- This is a good opportunity to ensure that good propulsion technique is being used.

Special considerations for manual wheelchairs operated by caregivers (Version 2)

- None.

Special considerations for powered wheelchairs operated by users (Version 3)

- *Adjustment tips*
  - ○ For the mode used for longer distances, the controller setting can be adjusted by the therapist or dealer to one that permits more speed and less sensitivity. Also, the deceleration distance should be increased so that a sudden stop does not cause the wheelchair user to fall or tip forward.
- The subject should alter the controller mode and speed settings to the ones most appropriate for the task.

Special considerations for powered wheelchairs operated by caregivers (Version 4)

- None.

Special considerations for scooters operated by users (Version 5)

- The scooter user can set the speed control so that he/she can proceed at the desired speed with the lever fully pushed or pushed partway depending upon the user's preference.
- The stiff suspension of most scooters can lead to some bouncing over rough surfaces such as sidewalk cracks.

7.20 Avoids moving obstacles

Versions applicable

Version	Wheelchair Type	Subject Type	Applicable
1	Manual	Wheelchair user	✓
2		Caregiver	✓
3	Powered	Wheelchair user	✓
4		Caregiver	✓
5	Scooter	Wheelchair user	✓

Skill level
- Intermediate.

Description
- While moving, the subject or learner avoids moving obstacles approaching from different directions.

Rationale
- In addition to stationary obstacles (dealt with in earlier skills), wheelchair operators must avoid moving obstacles (e.g., other wheelchair users, pedestrians) to avoid injury to themselves or others.

Prerequisites
- The "rolls forward short distance" skill is a prerequisite if this skill is carried out in the forward direction and the "rolls backward short distance" skill is a prerequisite if this skill is carried out in the backward direction.

Spotter considerations
- Spotter starting position: As for the "rolls longer distance" skill.
- Risks requiring spotter intervention: Forward or sideways tip or fall due to a sudden stop or turn.

Wheelchair Skills Test (WST)
Equipment
- Corridor or pathway as for the "rolls longer distance" skill. Although this skill is usually assessed with the "rolls longer distance" skill, a subject need not be able to pass the "rolls longer distance" skill to be tested for and pass the "avoids moving obstacles" skill. A shorter distance (e.g., 10 m) can provide the opportunity to assess this skill.

- An unoccupied manual wheelchair or equivalent for the tester to push. Although using his/her body as the moving obstacle is permitted, the tester is not expected to endanger him/herself.

Starting positions
- As for the "rolls longer distance" skill.
- Tester: The tester stands behind the unoccupied wheelchair that is being used as the moving obstacle, holding the push-handles, near the pathway but not in it. The tester should be able to see the approaching subject.

Instructions to subject
- *"Avoid bumping into anyone or anything that gets in your way."*
- The tester waits until the wheelchair gets close. Then, moving at a normal walking speed, the tester pushes the unoccupied wheelchair forward at a right angle into the path of the subject's wheelchair and stops (Figure 7.58). The tester times his/her movement to provide the subject with 2–3 seconds to avoid a collision. The distance away when the tester begins to move will need to be greater if the subject is moving quickly. This moving obstacle challenge is done again later, from the other side. If a collision appears to be imminent, the tester should take evasive action.

Capacity criteria
- As for the general scoring criteria, with the clarifications below.
- A "pass" should be awarded if:
  - The subject avoids any contact with the moving obstacle (Figure 7.59) without the tester needing to take evasive action.
  - The subject may avoid contact by stopping, slowing down, and/or changing direction.

Figure 7.58 For the "avoids moving obstacles" skill, the tester approaching at right angles.

Figure 7.59 To avoid the moving obstacle, the wheelchair user steers away.

- A "pass with difficulty" should be awarded if the subject has minimal but insignificant contact (i.e., insufficient to potentially cause injury to the wheelchair occupant or another person).
- A "fail" score should be awarded if the subject fails the "rolls forward short distance" skill.

Special considerations for manual wheelchairs operated by users (Version 1)
- None.

Special considerations for manual wheelchairs operated by caregivers (Version 2)
- None.

Special considerations for powered wheelchairs operated by users (Version 3)
- None.

Special considerations for powered wheelchairs operated by caregivers (Version 4)
- None.

Special considerations for scooters operated by users (Version 5)
- None.

Wheelchair Skills Training

General training tips
- This skill builds on the earlier skills that involve stopping and turning.
- The subject should be alert to the moving environment while the wheelchair is moving.
- If a hallway is clear enough to permit it, it may be advisable to drive in the middle of the hallway, to avoid collisions with people unexpectedly coming around corners or out of doors.
- The subject should obey driving conventions (the "rules of the road"), with respect to altering course to one side (to the right in North America)

when approaching others, use of horn or verbal warnings, overtaking and slowing down when approaching others or blind intersections.
- Sudden stops or changes of direction can lead to the wheelchair user falling forward or to the side in the wheelchair.

- *Progression*
  - The subject should start with a single moving obstacle moving slowly at a consistent speed, seen well in advance, to obstacles moving more rapidly and unpredictably, with less warning (e.g., actual pedestrian traffic in a crowded setting).
  - The subject should start with moving obstacles that approach from right angles and progress to ones coming from different angles, including overtaking and being overtaken.
  - The subject should start slowly and progressively increase the speed of propulsion.

- *Variations*
  - Different moving obstacles can be used (e.g., a rolled ball, a swinging pendulum).

Special considerations for manual wheelchairs operated by users (Version 1)

- *Progression*
  - Sudden stops can transfer weight forward onto the casters, allowing the unloaded rear wheels to skid.
  - The wheelchair user can practice both quick stops (leaning back and grabbing both hand-rims firmly) and swerves (leaning toward the direction of turn and grabbing one hand-rim firmly).
  - Some highly skilled wheelchair users can induce a controlled wheelie by throwing the trunk backward while coasting quickly forward. The goal is to overshoot the balance point and then grasp the hand-rims firmly to stop the wheelchair and prevent a rear tip. With a different amount of force applied to the two hand-rims, a rapid turn can be made.

Special considerations for manual wheelchairs operated by caregivers (Version 2)

- *Adjustment tip*
  - Secure push-handles are important for this skill.
- Sudden changes in speed or direction can cause the wheelchair user to fall forward or to the side. The caregiver should use good spotting techniques, reaching forward or to the side with a hand to stabilize the wheelchair user.

Special considerations for powered wheelchairs operated by users (Version 3)

- *Adjustment tip*
  - Adjusting the deceleration settings at top speeds is important for this skill. However, the higher the deceleration distance, the more planning is required to avoid the obstacle.

Special considerations for powered wheelchairs operated by caregivers (Version 4)

- None.

Special considerations for scooters operated by users (Version 5)

- The high speed that is possible with some scooters, combined with the high center of gravity and narrow wheelbase can make the scooter vulnerable to sideways tips during sudden turns.

7.21 Ascends slight incline

Versions applicable

Version	Wheelchair Type	Subject Type	Applicable
1	Manual	Wheelchair user	✓
2		Caregiver	✓
3	Powered	Wheelchair user	✓
4		Caregiver	✓
5	Scooter	Wheelchair user	✓

Skill level
- Intermediate.

Description
- The subject moves the wheelchair from a level surface up a slight incline to another level surface.

Rationale
- Inclines are encountered frequently in the natural and built environments. For instance, a 5° (~1:12) grade meets the current building codes for ramps in North America.

Prerequisites
- None.

Spotter considerations
- Spotter starting position: Behind the wheelchair, holding onto the spotter strap (if a manual wheelchair).
- Risks requiring spotter intervention if moving forward up the incline:
  - Rear tip when initially accelerating.
  - Forward tip or fall due to deceleration when striking the lower floor-incline transition.
  - Hyper-flexion injury of the lower limb at the lower floor-incline transition.
  - Rear tip while ascending the incline.

Wheelchair Skills Test (WST)

Equipment
- Incline at least 2.5 m long and at least 1.5 m wide.
- A lip and a handrail on both sides of the incline are desirable to prevent injuries but handrails should not be used in the performance of the skill.
- The incline should end at the upper end on a level surface or platform that is large enough for wheelchairs of all types, caregivers, and WST

personnel to turn around safely (2.0 m² or more is recommended). A lip around the open edges of the platform is recommended.
• There should be little or no lip at the lower junction of the floor and incline. The ability to overcome such obstacles is tested elsewhere.

Starting positions
• Wheelchair: On the level at the bottom of the incline, with the leading wheels of the wheelchair facing the incline and at least 0.5 m away. Some subjects may need to start farther away if they need to use momentum to get up the ramp. This is the subject's choice but the tester should not suggest this solution.

Instructions to subject
• *"Move the wheelchair up the ramp, without using the ramp handrails."*

Capacity criteria
• As for the general scoring criteria, with the clarifications below.
• A "pass" score should be awarded if:
  o The subject may use any type of propulsion, in the forward or backward direction.
  o All wheelchair parts are completely off the incline at the top.
  o The handrails are not grasped, and no wheel goes outside the lateral boundaries of the incline, although the subject or wheelchair may make contact with the ramp lips or rails without penalty.
• A "pass with difficulty" should be awarded if:
  o The footrests or RADs make enough contact with the surface at the lower transition (Figure 7.60) to significantly interfere with progression.

Figure 7.60 At the transition from a level surface to an incline, the RADs of a powered wheelchair contact the level surface.

    o A significant transient wheelchair tip occurs.

    o A foot catches on the floor as the wheelchair continues to move forward, without injury.

    o The subject exhibits overexertion symptoms due to unaccustomed exercise.

    o The thumb is injured to a minor degree by the wheel locks during forward hand thrusts.

Special considerations for manual wheelchairs operated by users (Version 1)

- The subject may use any type of propulsion (e.g., arm and leg, feet only, forward, backward).

Special considerations for manual wheelchairs operated by caregivers (Version 2)

- None.

Special considerations for powered wheelchairs operated by users (Version 3)

- None.

Special considerations for powered wheelchairs operated by caregivers (Version 4)

- None.

Special considerations for scooters operated by users (Version 5)

- None.

Wheelchair Skills Training

General training tips

- Some of these training tips also apply to incline descent and to inclines of different degrees.
- The steeper the incline, the greater is the likelihood of problems due to scraping the footrests or anti-tip devices at the transition between the lower end of the incline and the level landing area.
- If the drive wheels are uphill, they become relatively unloaded. This can cause loss of traction so that propulsion, braking, and directional control become difficult. If traction is lost to the extent that the wheels spin or the wheelchair begins to slide, the wheelchair user should lean toward the affected wheels. If this is insufficient, then the wheelchair should be turned around so that the drive wheels are downhill. It is best to turn around on the level but, if that is not possible, the wheelchair user should lean uphill during the turn.
- Edges and drop-offs at the sides of the incline or at the sides of the platform at the top of the incline should be avoided to prevent injury.
- Momentum can be used to ascend short inclines by approaching at speed. However, if the wheelchair user strikes the floor-ramp transition too quickly, he/she may tip the wheelchair forward or fall forward out of the wheelchair.

- *Progression*
  - o The subject should start with the wheelchair stationary at the lower end of the incline and progress to a moving approach.
  - o The subject should start with a minimal incline and proceed to more extreme ones.

- *Variations*
  - o Inclines with different surfaces, such as grass, cobblestone, or loose rock.
  - o Stopping and steering on the incline.
  - o If a ramp is wide enough, the wheelchair user can steer back and forth across the incline ("slalom" or "zig-zag"), to decrease the apparent slope (Figure 7.61). Pylons can be set up to provide a path for the wheelchair to follow. The more turns used, the lower the effective slope (but the greater the distance travelled). Although a slalom path up a steep incline will reduce the effective slope, it will introduce an element of side-slope (dealt with later in the "rolls across side-slope" skill).

Special considerations for manual wheelchairs operated by users (Version 1)

- *Adjustment tips*
  - o A heavy knapsack will reduce rear stability. It can be moved to the lap (although this may limit forward lean) or footrests.
  - o At the lower transition, either ascending or descending, the clearance of footrests can create problems.
  - o If the RADs are too low, this can cause rear-wheel "float" whereby the rear wheels are not in contact with the surface (because the

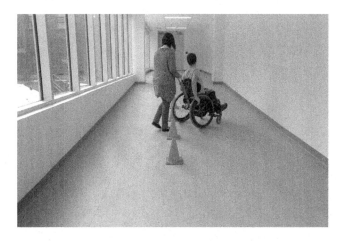

Figure 7.61 Performing a slalom through pylons while ascending a slight incline.

189

wheelchair is suspended between the casters and anti-tip devices) and are therefore unable to be used for propulsion or braking.

o "Grade aids" (or "hill holders") may be used. These are attachments that, when activated, allow the rear wheels to roll forward but not backward. These devices allow the wheelchair user to rest on the incline without rolling back. The wheelchair user should apply them before he/she starts up the incline.

o Some wheelchairs have gears that permit inclines to be handled more easily.

- *Two-hand-propulsion pattern*
  o When negotiating the incline-floor transition at the lower end, during either ascent or descent, the wheelchair user should be careful not to catch an unsupported foot, as this could lead to a hyper-flexion injury of the knee.
  o When getting the casters onto the bottom of an incline, it may be necessary to transiently tip the wheelchair ("popping" the casters) if the footrests are low and to reduce the sudden braking that occurs at the transition.
  o Some wheelchair users use a rocking action to get the casters over the initial lip (if any).
  o The wheelchair user should lean forward as he/she goes up the ramp to apply more force to the hand-rims and to avoid tipping backward (Figure 7.62). The need for forward lean increases as the slope increases. In addition to a consistent forward lean, it can be helpful to lean forward a little more with each push to apply greater forces to the hand-rims.

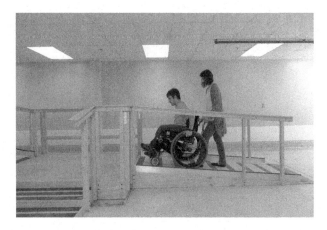

Figure 7.62 Leaning forward while ascending a slight incline.

190

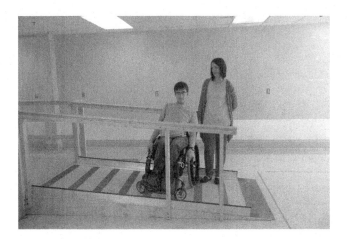

Figure 7.63 Resting turned sideways across an incline.

- o If the wheel locks are not of the retractable type, forward leaning can result in injury to the backs of the thumbs if the wheelchair user is not careful.
- o It may be necessary to use shorter propulsive strokes than on the level, to avoid rolling backward between strokes.
- o The recovery path of the hands at the end of the propulsive stroke may be more like an arc (following the hand-rim) than a loop (below the hand-rim) for this skill.
- o If the wheelchair user gets tired, part of the way up the incline, he/ she should turn the wheelchair to the side and rest (Figure 7.63). This can be done without applying the wheel locks.
- o If the wheelchair starts to roll backward, instead of grasping both hand-rims (that might cause a rear tip), the wheelchair user can grab one. As the other wheel rolls backward, this will turn the wheelchair across the slope.
- o *Variations*
  - – Alternating hands during propulsion may help to prevent rollback.
  - – If using a slalom path up the incline, the wheelchair user will generally turn uphill (e.g., 90°) at the end of each traverse to go back the other way. However, if this is not possible due to limitations of strength or stability, the turn may be downhill (e.g., 270°). Although a little height up the incline is lost, the additional speed during the turn provides momentum to assist in regaining the loss.

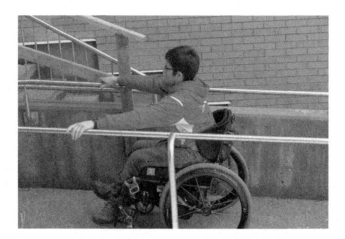

Figure 7.64 Using ramp handrails to ascend an incline.

- As a learning exercise, it may be helpful to have the wheelchair user try to ascend the incline (with a spotter) without leaning forward.
- The wheelchair user may use the ramp handrails if available (Figure 7.64).

- *Hemiplegic-propulsion pattern*
  - It is usually easier for a wheelchair user with hemiplegia (who propels the wheelchair with one arm and one leg) to go up the incline backward (Figure 7.65). Whenever rolling resistance is encountered (including when ascending inclines), foot propellers find it easier to push backward than to pull forward with the feet.

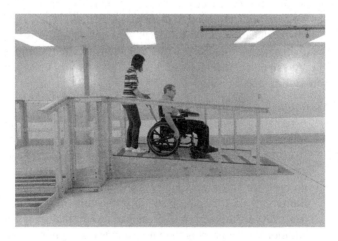

Figure 7.65 Person with hemiplegia ascending an incline backward.

Figure 7.66 Both feet on the same footrest.

Special considerations for manual wheelchairs operated by caregivers (Version 2)

- If the wheelchair user has hemiplegia, the caregiver can put the strong-side foot on the other footrest to avoid it getting caught on the transition (Figure 7.66).
- To push the wheelchair forward up an incline, the caregiver should bend his/her knees and lean toward the wheelchair. The caregiver should not use his/her knee to apply pressure to the backrest.

Special considerations for powered wheelchairs operated by users (Version 3)

- A small lip on the side of an incline may be sufficient to prevent a manual wheelchair from accidentally going over the edge, but a powered wheelchair can go over such a lip more easily.
- Most powered wheelchairs can handle 5° with ease, at least from the perspective of having enough power to manage the slope.
- Altering the position of the wheelchair seat (i.e., with respect to tilt, recline, seat height) may be helpful to improve stability or alter the weight distribution on the wheels (e.g., for more traction).
- The tilt or leg-elevation functions can be used to avoid scraping the footrests at the lower incline transition.

- *Progression*
  - o The subject may begin training with the controller in a low setting but programming that provides more power and torque may be needed for success. The user may need to change to a different drive mode to get up the incline.

193

Special considerations for powered wheelchairs operated by caregivers (Version 4)
- If the space is narrow and the caregiver must operate the wheelchair from in front, the caregiver should be careful not to run over his/her own toes.
- If the wheelchair does not have a headrest and if the wheelchair user is having difficulty in maintaining an upright head position while ascending an incline, the caregiver can support the head with a hand.

Special considerations for scooters operated by users (Version 5)
- None.

7.22 Descends slight incline

Versions applicable

Version	Wheelchair Type	Subject Type	Applicable
1	Manual	Wheelchair user	✓
2		Caregiver	✓
3	Powered	Wheelchair user	✓
4		Caregiver	✓
5	Scooter	Wheelchair user	✓

Skill level
- Intermediate.

Description
- The subject moves the wheelchair from a level surface down a slight incline to another level surface.

Rationale
- As for the "ascends slight incline" skill.

Prerequisites
- None.

Spotter considerations
- Spotter starting position: If the wheelchair is to move forward down the incline, the spotter should be behind the wheelchair, holding onto the spotter strap (if a manual wheelchair) with one hand and the other hand in front of the wheelchair user's shoulder. If using two spotters, one spotter should be behind the wheelchair, holding onto the spotter strap and the second spotter should be in front of and beside the wheelchair to resist a forward tip or fall. A removable seat belt may be used if there is concern about the subject falling forward from the wheelchair.
- Risks requiring spotter intervention:
  - Rear tip if performed in the wheelie position.
  - Forward tip or fall due to deceleration when striking the lower incline-floor transition.
  - Hyper-flexion injury of the lower limb at the lower incline-floor transition.
  - Runaway leading to collision or tip-over.
  - Hand injuries to the wheelchair user due to friction burns or lacerations from hand-rim irregularities if the wheelchair is allowed to descend too rapidly.

o Thumb injury on the wheel locks if the wheelchair user grabs the hand-rims when they are rolling too quickly because the hands can get pulled forward into the wheel locks.

Wheelchair Skills Test (WST)

Equipment
- As for the "ascends slight incline" skill.

Starting positions
- Wheelchair: All wheels are on the level surface at the top of the incline with the leading wheels of the wheelchair facing the incline and at least 0.5 m away.

Instructions to subject
- *"Move the wheelchair down the ramp, without using the ramp hand-rails. Keep the wheelchair under control."*

Capacity criteria
- As for the general scoring criteria, with the clarifications below.
- A "pass" should be awarded if:
  - o All wheelchair parts are completely off the incline at the bottom.
  - o The handrails are not grasped and no wheel goes outside the lateral boundaries of the incline, although the subject or wheelchair may make contact with the ramp lips or rails without penalty.
  - o The subject is under control during the full descent, including the transition to level ground.
- A "pass with difficulty" should be awarded if the subject catches the foot on the floor as the wheelchair continues to move forward, without injury.

Special considerations for manual wheelchairs operated by users (Version 1)
- A "pass" should be awarded if:
  - o The subject may use any type of propulsion (e.g., arm and leg, feet only, forward, backward).
  - o The wheelie position may be used for descending all or part of the incline.
  - o It is permissible for the subject to use the bottoms of the shod feet as wheel locks.
- A "pass with difficulty" should be awarded if:
  - o The subject sustains mild friction burns of the hands.
  - o There is minor thumb injury on the wheel locks because the hands get pulled forward into the wheel locks.
  - o The subject drags the bottoms of unshod feet to slow the wheelchair by friction between the feet and floor.

o The subject drags the toes, even if the feet are shod.
• Comment only: The subject uses the wheel locks as rolling wheel locks (e.g., by partially or repeatedly applying them).

Special considerations for manual wheelchairs operated by caregivers (Version 2)
• None.

Special considerations for powered wheelchairs operated by users (Version 3)
• Disengaging the motors and letting the wheelchair roll down the ramp is not considered a safe method.

Special considerations for powered wheelchairs operated by caregivers (Version 4)
• None.

Special considerations for scooters operated by users (Version 5)
• None.

Wheelchair Skills Training

General training tips
• A smooth straight controlled descent in the forward direction is the basic method.
• The subject should proceed slowly to maintain control and should be prepared to stop at any time. It is easier to maintain speed control than to regain it after it has been lost.
• There are many similarities to the "ascends slight incline" skill.

Special considerations for manual wheelchairs operated by users (Version 1)

• *Two-hand-propulsion pattern*

  o *Adjustment tip*
    – Appropriate and controllable friction between the hands and hand-rims is important to carry out this skill safely and effectively. Gloves are helpful. The type of coating (if any) on the hand-rims affects friction, as do hand-rim size and shape. A quick and inexpensive way to increase the friction of a hand-rim is to spiral wrap it with rubber tubing.
  o The wheelchair user should keep his/her weight back (Figure 7.67), to maintain good traction on the rear wheels and to avoid forward tips or falls.
  o To slow down or steer, the wheelchair user should hold the hands still at the 1:00 o'clock position and let the hand-rims slide through his/her fingers. It is generally better to provide continuous friction than to use a jerky grasp-and-release method. However, the

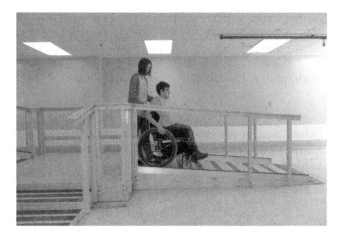

Figure 7.67 Leaning back while descending an incline.

grasp-and-release method may be useful to minimize the heat that builds up through friction, grasping either with both hands at the same time or alternating from one to the other.

o If the wheelchair starts to roll too quickly and the incline is wide enough, instead of grasping both hand-rims to stop, the wheelchair user can grab one, turning across the slope.

o *Variations*

– If a ramp is wide enough, the wheelchair user can slalom down it by letting the hand-rim of one wheel at a time slide through the fingers. By descending using the slalom method, the apparent slope of the incline is lessened. Also, this technique may prevent the hands from overheating due to sustained friction. To provide the subject with a clear indication of how turning can be controlled while descending an incline, the trainer can ask the subject to roll straight down an incline for a short distance, then grab one hand-rim—a dramatic turn will result. This can be repeated with more gradual applications of force on the hand-rim to progress toward smooth controlled turns in either direction. Downhill-turning tendency (see "gets across side-slope" skill later) can be used to advantage when the wheelchair user wishes to turn downhill. Leaning forward will accentuate the tendency and ease the turn.

– Caution should be used when using the wheel locks as moving wheel locks. This is not a commonly recommended method.

– The wheelchair user may use the handrails of the incline, if available.

- If the wheelchair user has weak trunk muscles and a tendency to fall forward when facing downhill on inclines, he/she may feel more comfortable descending the incline backward. The backward approach may also be used if, when descending forward on a steeper incline, the wheelchair user experiences loss of traction due to the unloading of the uphill wheels. When going downhill backward, the wheelchair user should lean uphill to reduce the likelihood of tipping over backward. As with any time the wheelchair is moving backward, it is important to proceed slowly with frequent shoulder checks and to avoid sudden stops that can cause rear tips.
    - Using a wheelie is effective but will be discussed later under the "descends steep incline in wheelie position" skill section.

- *Hemiplegic-propulsion pattern*
    - The wheelchair user can proceed forward down the incline, using the foot to slow down.
    - The wheelchair user needs to be cautious that the foot does not get caught under the chair at the lower incline-floor transition.

Special considerations for manual wheelchairs operated by caregivers (Version 2)

- *Adjustment tip*
    - The push-handles should be checked to ensure that they will not pull off.
- The basic method is in the forward direction with all four wheels on the incline.
- The caregiver holds the push-handles firmly and allows the wheelchair to roll down the ramp while controlling the speed.
- The caregiver avoids sudden stops and slows down as he/she reaches the bottom transition to level ground.
- The caregiver can put one hand on the wheelchair user's shoulder to prevent a forward fall and also to steer the wheelchair as the wheelchair will tend to twist if only one push-handle is held.

- *Variations*
    - The forward descent can be performed in the wheelie position. This is useful on steep inclines, to prevent the wheelchair user from falling forward. However, this method may require the caregiver to bend too far forward, which may strain the back.
    - Another method is to descend backward. This ensures that the wheelchair does not run away from the caregiver and that the wheelchair user does not fall forward. The caregiver should look over the shoulder for obstacles.

Special considerations for powered wheelchairs operated by users (Version 3)

- Altering the position of the wheelchair seat (i.e., with respect to tilt, recline, seat height) may be helpful to improve stability, alter the weight distribution on the wheels (e.g., for more traction), or ensure footrest clearance at the lower transition. However, some wheelchairs do not permit the wheelchair to be driven when the positioning options exceed a threshold.
- Training should begin with the controller in a low setting.
- In a powered wheelchair, unlike a two-hand-propelled manual one, only one hand is needed to control speed and direction. The other arm can be hooked around the backrest or push-handle to prevent falling forward onto the lap.

Special considerations for powered wheelchairs operated by caregivers (Version 4)

- If the space is narrow and the caregiver must operate the wheelchair from in front, the caregiver should be careful not to run over his/her own toes.

Special considerations for scooters operated by users (Version 5)

- None.

7.23 Ascends steep incline

Versions applicable

Version	Wheelchair Type	Subject Type	Applicable
1	Manual	Wheelchair user	✓
2		Caregiver	✓
3	Powered	Wheelchair user	✓
4		Caregiver	✓
5	Scooter	Wheelchair user	✓

Skill level
• Advanced.

Description
• The subject moves the wheelchair from a level surface up a steep incline (Figure 7.68) to another level surface.

Rationale
• Inclines with slopes greater than the standard recommended value are encountered frequently in the natural and built environments. The appropriate technique for a steep incline may differ somewhat from that used for a lesser slope.

Prerequisites
• "Ascends slight incline" skill.

Spotter considerations
• As for "ascends slight incline" skill.

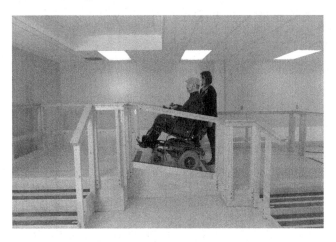

Figure 7.68 Ascending a steep incline.

201

Wheelchair Skills Test (WST)

Equipment
- As for "ascends slight incline" skill, except that the incline has a 10° slope.

Starting positions
- As for "ascends slight incline" skill.

Instructions to subject
- *"Move the wheelchair up the ramp, without using the ramp handrails."*

Capacity criteria
- As for the general scoring criteria, with the clarifications below.
- As for "ascends slight incline" skill.
- A "fail" score should be awarded if the subject fails the "ascends slight incline" skill.

Special considerations for manual wheelchairs operated by users (Version 1)
- None.

Special considerations for manual wheelchairs operated by caregivers (Version 2)
- None.

Special considerations for powered wheelchairs operated by users (Version 3)
- None.

Special considerations for powered wheelchairs operated by caregivers (Version 4)
- None.

Special considerations for scooters operated by users (Version 5)
- None.

Wheelchair Skills Training

General training tips
- As for the "ascends slight incline" skill.

- *Progression*
  - Although only 5° and 10° inclines are mentioned specifically in this book, for learners and wheelchairs capable of handling steeper inclines, it is reasonable to attempt these under the supervision of a trainer, even if only to help the subject recognize the limits of what is possible for him/her with that wheelchair.

Special considerations for manual wheelchairs operated by users (Version 1)
- As for the "ascends slight incline" skill.
- As the steepness of the incline increases, the wheelchair user leans further forward and the initial contact with the hand-rims moves forward. The propulsion contact angle diminishes (although the duration of the push phase remains similar) and the force increases. The recovery phase becomes faster and an arc recovery pattern (back along the hand-rims) may be used.
- For very steep inclines, some wheelchair users will go up backward in the wheelie position. This requires considerable skill and strength. The uphill movement is initiated by allowing the wheelchair to fall ("dip") partially backward, followed by a strong pull backward on the hand-rims to re-achieve balance a short distance up the slope.

Special considerations for manual wheelchairs operated by caregivers (Version 2)
- As for the "ascends slight incline" skill.

Special considerations for powered wheelchairs operated by users (Version 3)
- As for the "ascends slight incline" skill.

Special considerations for powered wheelchairs operated by caregivers (Version 4)
- As for the "ascends slight incline" skill.

Special considerations for scooters operated by users (Version 5)
- As for the "ascends slight incline" skill.
- Most scooters have adequate power to get up even steeper inclines.
- Scooters may have difficulty at the upper incline-level transition due to inadequate clearance ("break-under angle") between the front and back wheels.
- Scooters may have difficulties at the lower incline-level transition if any rigid RADs cause the rear wheels to "float" off the surface. Approaching with a little extra speed may help, but the stiff suspension of many scooters may cause the scooter user to bounce off the seat causing a loss of control.

7.24 Descends steep incline

Versions applicable

Version	Wheelchair Type	Subject Type	Applicable
1	Manual	Wheelchair user	✓
2		Caregiver	✓
3	Powered	Wheelchair user	✓
4		Caregiver	✓
5	Scooter	Wheelchair user	✓

Skill level
- Advanced.

Description
- The subject moves the wheelchair from a level surface down a steep incline (Figure 7.69) to another level surface.

Rationale
- As for the "ascends steep incline" skill.

Prerequisites
- "Descends slight incline" skill.

Spotter considerations
- As for "descends slight incline" skill.

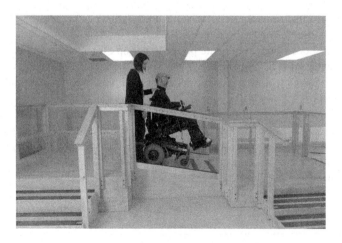

Figure 7.69 Descending a steep incline.

Wheelchair Skills Test (WST)

Equipment
- As for "ascends steep incline" skill.

Starting positions
- As for "descends slight incline" skill.

Instructions to subject
- As for "descends slight incline" skill.

Capacity criteria
- As for the general scoring criteria, with the clarifications below.
- As for "descends slight incline" skill.
- A "fail" score should be awarded if the subject fails the "descends slight incline" skill.

Special considerations for manual wheelchairs operated by users (Version 1)
- None.

Special considerations for manual wheelchairs operated by caregivers (Version 2)
- None.

Special considerations for powered wheelchairs operated by users (Version 3)
- None.

Special considerations for powered wheelchairs operated by caregivers (Version 4)
- None.

Special considerations for scooters operated by users (Version 5)
- None.

Wheelchair Skills Training

General training tips
- As for the "descends slight incline" skill.

Special considerations for manual wheelchairs operated by users (Version 1)
- As for the "descends slight incline" skill.
- The descent in the wheelie position is dealt with in a later skill.

Special considerations for manual wheelchairs operated by caregivers (Version 2)
- As for the "descends slight incline" skill.

Special considerations for powered wheelchairs operated by users (Version 3)
- As for the "descends slight incline" skill.
- When stopping while descending a steep incline, moving the joystick into reverse or turning the power off may work better than simply bringing the joystick to the resting neutral position.

Special considerations for powered wheelchairs operated by caregivers (Version 4)
- As for the "descends slight incline" skill.

Special considerations for scooters operated by users (Version 5)
- As for the "descends slight incline" skill.

7.25 Rolls across side-slope

Versions applicable

Version	Wheelchair Type	Subject Type	Applicable
1	Manual	Wheelchair user	✓
2		Caregiver	✓
3	Powered	Wheelchair user	✓
4		Caregiver	✓
5	Scooter	Wheelchair user	✓

Skill level

- Intermediate.

Description

- The subject moves the wheelchair across a slight side-slope (Figure 7.70) without turning downhill or uphill significantly, then repeats the task in the opposite direction.

Rationale

- Side-slopes (or cross-slopes) are frequently encountered in built and natural environments. Sidewalks, for instance, are usually sloped 2% (1:50) toward the street to allow water to run off. Steeper grades are also often found (e.g., where sidewalks cross driveways). The yaw axis of a wheelchair (i.e., turning toward the left or right) is between the drive wheels. If the combined center of gravity of the wheelchair and user is ahead of the drive wheels (as is usually the case with rear-wheel drive

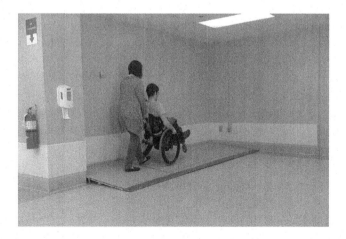

Figure 7.70 Crossing a side slope in the wheelie position.

wheelchairs), the wheelchair will tend to turn downhill on a side slope (downhill-turning tendency). If the combined center of gravity of the wheelchair and user is behind the drive wheels (as is usually the case with front-wheel-drive wheelchairs), the wheelchair will tend to turn uphill on a side slope (uphill-turning tendency).

Prerequisites
- None.

Spotter considerations
- Spotter starting position: Slightly behind and downhill from the wheelchair.
- Risks requiring spotter intervention: Sideways tip or fall downhill.

Wheelchair Skills Test (WST)
Equipment
- Incline of 5°, at least 2 m long (in the line of progression) and at least 1.5 m wide.
- At least an extra 1.5 m before the starting line and beyond the finishing line.
- Start and finish lines perpendicular to the line of progression.
- Means of monitoring if any of the downhill wheels drift or turn downhill by greater than 10 cm from the starting position. The slope-level transition can be used, or any line parallel to it.

Starting positions
- Wheelchair: With the wheel locks off, and all wheels on the sloped surface, oriented in the line of progression across the slope. The downhill drive wheel is positioned 10 cm up the slope from the line used to detect whether the wheelchair has turned or drifted downhill. The axles of the leading wheels must be behind the starting line. The casters should be trailing appropriately for the direction of travel so that there is no initial deflection of the wheelchair due to the casters realigning themselves.

Instructions to subject
- *"Move the wheelchair across the slope to the finish line* (indicate it) *without letting the wheels turn downhill below the line* (indicate it)."
- *"Now do the same thing in the other direction."*

Capacity criteria
- As for the general scoring criteria, with the clarifications below.
- A "pass" should be awarded if:
  o The leading wheels cross the finish line.
  o Any path may be used as long as no downhill wheel crosses the line 10 cm downhill from the starting position.

Special considerations for manual wheelchairs operated by users (Version 1)
- None.

Special considerations for manual wheelchairs operated by caregivers (Version 2)
- The caregiver's feet need not remain above the line being avoided because the caregiver's usual position relative to the wheelchair is slightly downhill to the wheelchair.

Special considerations for powered wheelchairs operated by users (Version 3)
- A front-wheel-drive wheelchair will tend to self-steer uphill instead of downhill, but there is no penalty for this if the wheelchair is able to complete the 2 m in the space available without any downhill wheel (e.g., a rear caster) in contact with the surface moving below the line.

Special considerations for powered wheelchairs operated by caregivers (Version 4)
- None.

Special considerations for scooters operated by users (Version 5)
- None.

Wheelchair Skills Training

General training tips
- The extent of downhill- or uphill-turning tendency is directly proportional to how far the combined center of gravity of the wheelchair and occupant is in front of or behind the drive wheels. The person operating the wheelchair can take steps to minimize this distance by repositioning the center of gravity (e.g., by leaning, tilting, reclining).
- If there is room to do so on a path, the person operating the wheelchair should stay away from the downhill edge of a side slope to avoid veering off the path.

- *Variation*
  - o Slowly turning the wheelchair 360° in place on a side slope will provide a good sense of how downhill-turning tendency affects the wheelchair at different angles.

- *Progression*
  - o Although only a 5° side slope is mentioned specifically in this book, for subjects and wheelchairs capable of handling steeper inclines, it is reasonable to attempt these under the supervision of a trainer, even if only to help the subject recognize the limits of what is possible for him/her with that wheelchair.

Special considerations for manual wheelchairs operated by users (Version 1)

- *Adjustment tip*
  - Moving the rear axles of a rear-wheel-drive wheelchair forward reduces the downhill-turning tendency.
- Side slopes require significantly more energy to push across.
- The wheelchair user should lean backward to keep the weight away from the casters.
- Although downhill-turning tendency can make it difficult to proceed in a straight line across a side slope, if the subject leans forward appropriately, this tendency can be used to facilitate turns on inclines.

- *Two-hand-propulsion pattern*
  - To avoid turning downhill, the wheelchair user should push harder on the downhill wheel.
  - Different push frequencies may be used for the two hands. For instance, when moving across a side-slope with the right side downhill, the right hand may push two to three times for every one push on the left.
  - When pushing longer distances, route planning can be used to avoid overuse on one side. For instance, part of the journey can be carried out on the right-hand sidewalk (where the left side is downhill) and part of the journey on the left-hand sidewalk.
  - In some cases, the uphill hand may be used exclusively for braking (to minimize downhill-turning tendency) rather than for assisting with propulsion.
  - Shorter stokes may need to be used to keep the wheelchair moving straight.
  - On steep cross-slopes, problems (e.g., loss of uphill-wheel traction, lateral tip over, folding of the wheelchair) may arise due to the lack of weight on the uphill wheel. These problems can be minimized by leaning uphill.
  - As noted earlier under the "descends slight incline" skill, downhill-turning tendency can be used to advantage when the wheelchair user wishes to turn downhill. Leaning forward will accentuate the tendency and ease the turn.

  - *Variations*
    - A useful learning experience to demonstrate the downhill-turning tendency is to have the wheelchair user lean forward, to illustrate how the downhill-turning tendency increases.
    - If there is an uphill wall that can be used, the wheelchair user can drag the uphill hand on the wall behind the rear axle to counteract the downhill-turning tendency. This is analogous to the drag turn discussed earlier.

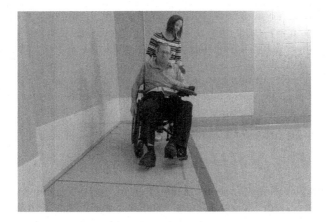

Figure 7.71 Drifting downhill on a side slope with the sound-side uphill.

- In the wheelie position facing across a slope, there is no down-hill-turning tendency, because the center of gravity is between the rear wheels.

- *Hemiplegic-propulsion pattern*
  - Some users may choose to go backward with the sound-side down-hill rather than forward with the sound-side uphill (Figure 7.71), to help manage the downhill-turning tendency.
  - When learning the skill in the forward direction, it may be less frustrating to cross the side slope with the sound-side downhill first; this will tend to counteract rather than aggravate the downhill-turning tendency.

Special considerations for manual wheelchairs operated by caregivers (Version 2)
- To resist the downhill-turning tendency while pushing the wheelchair across a side slope, the caregiver needs to push harder on the downhill push-handle and pull back on the uphill push-handle.
- For a steeper slope, the caregiver may choose to use the wheelie position.
- If the wheelchair user is in a tilt-in-space or reclining wheelchair, tilting or reclining the wheelchair can be used to get the center of gravity farther back.

Special considerations for powered wheelchairs operated by users (Version 3)
- Although a rear-wheel-drive wheelchair will tend to turn downhill (analogous to a manual wheelchair), a front-wheel-drive wheelchair will tend to turn uphill.
- Many powered wheelchairs are equipped with automatic correction of downhill/uphill-turning tendency on side-slopes.

- If there is no automatic correction, the wheelchair user should aim slightly away from the expected deviation (i.e., aim uphill for a rear-wheel-drive wheelchair and downhill for a front-wheel-drive wheelchair).
- If the wheelchair user is in a tilt-in-space or reclining wheelchair, tilting or reclining the wheelchair can be used to get the center of gravity over the drive wheels.

Special considerations for powered wheelchairs operated by caregivers (Version 4)
- None.

Special considerations for scooters operated by users (Version 5)
- On steeper side slopes, sideways tips are possible due to the relatively narrow base width and high center of gravity of some scooters.

7.26 Rolls on soft surface

Versions applicable

Version	Wheelchair Type	Subject Type	Applicable
1	Manual	Wheelchair user	✓
2		Caregiver	✓
3	Powered	Wheelchair user	✓
4		Caregiver	✓
5	Scooter	Wheelchair user	✓

Skill level
- Intermediate.

Description
- The subject moves the wheelchair a short distance on a soft surface.

Rationale
- There are many soft surfaces (e.g., carpet, dirt, grass, gravel, sand, snow) with increased rolling resistance. Propulsion is more difficult on such surfaces because the wheels tend to sink into the surface (Figure 7.72), especially wheels that are narrow or of small diameter.

Prerequisites
- None.

Spotter considerations
- Spotter starting position: Behind the wheelchair, holding onto the spotter strap with one hand (if a manual wheelchair).

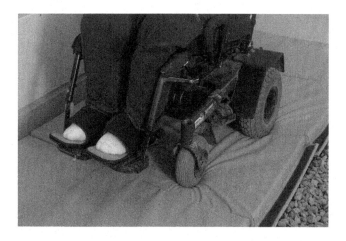

Figure 7.72 On a soft surface (gym mats), casters digging into the surface.

- Risks requiring spotter intervention:
  - ○ Rear tip when accelerating.
  - ○ Overuse injury due to the additional forces needed.

Wheelchair Skills Test (WST)

Equipment
- Pathway that includes a soft surface at least 2.0 m long and 1.5 m wide.
- There should be an additional 1.5 m of soft surface before the starting line and 1.5 m beyond the finishing line.
- Options for the soft surface include a gym mat (5 cm thick) (Figure 7.73), gravel (medium grade, 5–6 cm deep) (Figure 7.74), sand (fine grain,

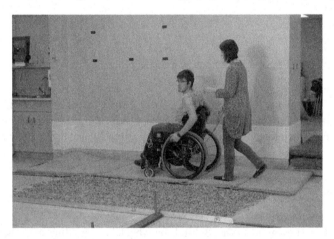

Figure 7.73 Using gym mats as a soft surface.

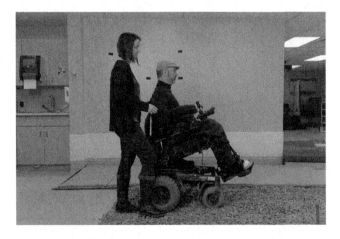

Figure 7.74 Using gravel as a soft surface.

214

5–6 cm deep), and indoor/outdoor carpet over 5 cm open-cell foam or equivalent.
- Note that some sand and gravel pits have lips that make it difficult to get into and out of them. It is the 2 m of soft surface that is the focus of this skill, not the entry and exit.

Starting positions
- Wheelchair: Fully on the soft surface with the leading wheel axles behind the starting line.

Instructions to subject
- "Move the wheelchair over the finish line (indicate it)."

Capacity criteria
- As for the general scoring criteria, with the clarifications below.
- A "pass" should be awarded if:
  - The leading wheel axles are beyond the finish line.
  - All techniques are permitted, such as forward or backward approaches.
  - During the course of any single attempt, a subject may use different approaches.

Special considerations for manual wheelchairs operated by users (Version 1)
- The wheelchair user may use the feet.
- Transiently popping the casters off the soft surface is an effective strategy.
- Rolling forward in the full wheelie position is also effective.

Special considerations for manual wheelchairs operated by caregivers (Version 2)
- None.

Special considerations for powered wheelchairs operated by users (Version 3)
- The wheelchair user may use the wheelchair's body positioning options (e.g., tilt, recline, leg-rest elevation) to reduce the weight on the smaller wheels.

Special considerations for powered wheelchairs operated by caregivers (Version 4)
- None.

Special considerations for scooters operated by users (Version 5)
- None.

Wheelchair Skills Training
General training tips

- *Adjustment tip*
  - The diameter, width, and shape of the wheels and tires will affect the extent to which they sink into the soft surface.

- When approaching a section of soft or irregular terrain, the wheelchair user should look ahead and plan a route that will minimize difficulties.
- When moving from a smooth level surface onto a soft surface, the wheelchair will decelerate, so it may be necessary to slow down (or pop the casters, if in a manual wheelchair) when approaching such a transition.
- To minimize rolling resistance, reducing the weight on the small wheels (casters) and increasing the weight on the drive wheels is a helpful strategy.
- When proceeding across a soft or rough surface, it is easiest to move forward in a straight line because, if the casters sink into the soft surface, they will be less free to swivel should the user wish to change direction.
- If one drive wheel is spinning, the wheelchair user should shift his/her weight in the direction of the slipping wheel to increase the traction.
- For rear-wheel-drive wheelchairs, it may be easier to lead with the larger wheels (i.e., in the backward direction). The larger-diameter wheels make it easier to get started. The casters will trail backward and the resulting longer wheelbase may help as well.

- *Variations*
  o A variety of surfaces (e.g., sand, thick carpet, foam, a gym mat, gravel) provide similar, but not identical, experiences.
  o If the surface is too soft to proceed over, a mat or other materials can be laid down over it. If an assistant is available, long distances can be covered by picking up the mat behind the wheelchair and moving it to the front, proceeding forward in a step-wise fashion.

Special considerations for manual wheelchairs operated by users (Version 1)

- *Adjustment tips*
  o This is the first of several skills for which it is beneficial to pop the casters off the surface or to perform a full wheelie. Any adjustment that lowers the rear stability of the wheelchair (e.g., moving the axles of the rear wheels forward) will make it easier to pop the casters.
  o It may be necessary to reposition the RADs to allow the wheelchair to be tipped backward sufficiently to transiently pop the casters (Figure 7.75). To reposition most RADs, the subject will need to press a button or release mechanism on the wheelchair frame that locks the RADs in place. The subject should note the position of the RADs, so that he/she will be able to restore them later. Then, the subject can either reposition the RADs so that they face upwards or remove them altogether. To restore the RADs, the subject should simply reverse the steps. Note that whenever the RADs have been inactivated, the wheelchair user is at increased risk of a rear tip. The spotter should be vigilant to spot the wheelchair user closely until

Figure 7.75 Repositioning the RADs to allow the wheelchair to be tipped backward for functional purposes.

the wheelchair user becomes used to this new condition. Even if the RADs are left in place, the wheelchair user should not rely on them to prevent rear tipping because they might sink into a soft surface.
o If the wheelchair has elevating footrests, it will be easier to pop the casters if the footrests are lowered.

- *Two-hand-propulsion pattern*
  o The forward approach to negotiating soft surfaces is preferred because the wheelchair user can see where he/she is going.
  o The wheelchair user should use long slow strokes to keep the wheels from slipping in loose surfaces.
  o Because there is more rolling resistance on soft surfaces, more force is required by the wheelchair user.
  o Leaning forward slightly may help the wheelchair user to apply more force to the hand-rims and to prevent the additional force from causing a rear tip. However, keeping as much weight as possible on the rear wheels (i.e., leaning backward) will improve traction and keep the front wheels from digging into the soft surface. The wheelchair user should experiment with the extent of trunk lean to find the optimum (the "sweet spot" between too much and too little).
  o As noted above under "adjustment tips," this is the first in a series of skills for which popping the casters off the surface is useful or necessary (Figure 7.76). Such caster pops are a good option for the wheelchair user, lifting the casters off the surface during each push, but letting them touch the surface as the hands recover for the next push.

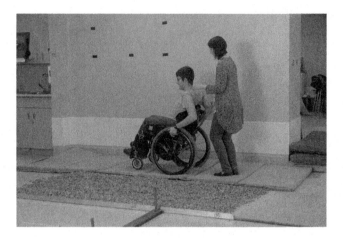

Figure 7.76 Transient caster pop to move forward on a soft surface.

○ During a caster pop, the longer the hands remain on the hand-rims, the farther forward the wheelchair will move with the casters off the surface. This can be thought of as analogous to taking a series of walking "steps" across the surface; a few long steps are preferable to many short steps.

○ *Progression*
  – For wheelchair users who are unfamiliar with caster pops, it can be a useful exercise to practice such pops on a smooth firm surface. The emphasis is on pushing the hand-rims forward but more forcefully than to simply roll forward and less forcefully than is needed to achieve a full wheelie position.

○ *Variations*
  – As a learning exercise, the wheelchair user may try the skill while leaning forward and backward to different extents, to find the optimum position for him/her.
  – If using the full wheelie position (a good option, but one that requires more skill) (Figure 7.77), the wheelchair user needs a strong forward "dip" to get going. If the casters touch the surface during the dip, the wheelchair user can lean forward slightly. This allows the casters to lift off further during the wheelie and provides better clearance during the dip.

• *Hemiplegic-propulsion pattern*
  ○ Rolling on soft surfaces with the hemiplegic-propulsion pattern (one arm and one leg) is easier in the backward direction, because there

218

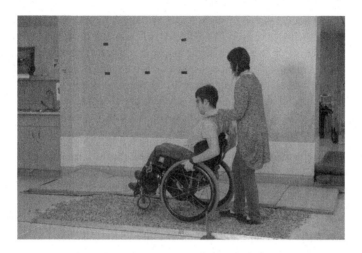

Figure 7.77 Using the full wheelie position to move forward on a soft surface.

is less rolling resistance with the large rear wheels than the smaller casters. Also when pushing backward with the foot, the casters become slightly unloaded which makes it easier to move them.

Special considerations for manual wheelchairs operated by caregivers (Version 2)
- To proceed in the forward direction, it may be necessary for the caregiver to lean forward to apply the extra force needed.
- The caregiver should not use his/her knee against the backrest of the wheelchair to apply more force because this may be uncomfortable for the wheelchair user (if the backrest is flexible) or dislodge a rigid removable backrest.

- *Variations*
  o The caregiver may find it easier to pull the wheelchair backward.
  o The caregiver may find it easier to tip the wheelchair back into the full wheelie position, so that almost all of the weight is on the rear wheels. The wheelchair can be pushed forward or pulled backward in the wheelie position. If there is very high rolling resistance, pulling may be more effective. However, pushing forward has the advantage that the caregiver can see where he/she is going. This is the first of many skills for which it may be useful for the caregiver to be able to transiently pop the casters from the ground briefly or to get into the full wheelie position. The caregiver should always let the wheelchair user know before he/she tips the wheelchair backward. To tip the wheelchair backward, the caregiver should use one foot on a tipping

219

lever (an extension of the wheelchair frame, to which the RAD may be attached) while pulling backward with the hands on the push-handles. For the full wheelie position, the caregiver should tip the wheelchair back far enough so that it is balanced over the rear wheels. How far back the chair needs to be tipped will vary depending on the wheelchair user and the wheelchair. To land after the assisted wheelie, the caregiver should slowly allow the casters to return to the floor using a foot on the tipping lever to help slow the landing.

Special considerations for powered wheelchairs operated by users (Version 3)

- If possible and necessary, the wheelchair user should adjust the controller setting to one that provides more torque.
- Positional control (e.g., tilt, recline) can alter the weight distribution between the front and rear wheels (Figure 7.78). It is easier to proceed on a soft surface if more of the weight is on wheels with larger diameter. Clearance for the feet can also be affected by this change.
- On soft or irregular terrain, there is an optimal speed that is fast enough to maintain forward movement but not so fast that the motion is uncomfortable or leads to a loss of control.
- Maintaining a steady speed is preferable to a series of stops and starts.
- On a "bottomless" soft surface (e.g., sand, gravel, mud), if the drive wheels are allowed to spin, the wheelchair may dig itself into a hole from which it can be difficult to get out without assistance.

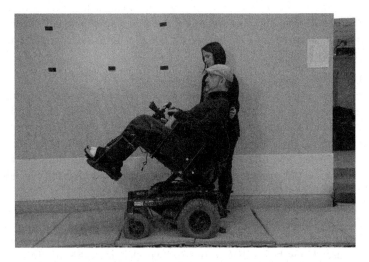

Figure 7.78 Tilting a powered wheelchair to unload the casters on a soft surface.

Special considerations for powered wheelchairs operated by caregivers (Version 4)

- With a rear-wheel-drive wheelchair, a caregiver can push down (or stand) on the back of the wheelchair to unload the casters and add traction to spinning wheels. The caregiver can also push forward, to assist with overcoming resistance. Alternatively, the front of the wheelchair can be lifted or pulled on. The converse is true for a front-wheel-drive wheelchair.

Special considerations for scooters operated by users (Version 5)

- None.

7.27 Gets over threshold

Versions applicable

Version	Wheelchair Type	Subject Type	Applicable
1	Manual	Wheelchair user	✓
2		Caregiver	✓
3	Powered	Wheelchair user	✓
4		Caregiver	✓
5	Scooter	Wheelchair user	✓

Skill level
- Intermediate.

Description
- The subject moves the wheelchair over a threshold.

Rationale
- Wheelchair users often encounter obstacles (e.g., door thresholds) that they may not be able to simply roll over. Alternative strategies may be needed. For example, a manual wheelchair user might need to pop the casters over the obstacle whereas a powered wheelchair user might need to change the mode setting to one with more power.

Prerequisites
- None.

Spotter considerations
- Spotter starting position: If using a single spotter, he/she should be behind the wheelchair, holding onto a spotter strap with one hand (if a manual wheelchair) and the other hand in front of the wheelchair user's shoulder. If using two spotters (as is recommended), the second spotter should stand to one side of the threshold. A removable seat belt can prevent the wheelchair user from falling from the wheelchair.
- Risks requiring spotter intervention:
  - Rear tip when accelerating to pop casters from the surface (if a manual wheelchair).
  - Forward tip or fall if the casters strike the threshold.

Wheelchair Skills Test (WST)

Equipment
- Threshold 2 cm high, 1.5 m wide, and 10 cm across (in the line of progression), rectangular in cross-section (i.e., a vertical front face without a bevel).
- The threshold should be secured so that it can withstand horizontal forces.

Starting positions
- Wheelchair: Facing the threshold with the leading wheels at least 0.5 m from it.

Instructions to subject
- *"Get your wheelchair over the threshold."*

Capacity criteria
- As for the general scoring criteria, with the clarifications below:
- A "pass" should be awarded if all parts of the wheelchair have passed beyond the threshold.
- A "pass with difficulty" should be awarded if:
  - There is significant jarring.
  - There is loss of control due to bouncing off the seat.
  - There is unintended hyper-flexion of the lower limb without injury.

Special considerations for manual wheelchairs operated by users (Version 1)
- The wheelchair user is permitted to use his/her feet or stand to get over the threshold.

Special considerations for manual wheelchairs operated by caregivers (Version 2)
- The caregiver may request assistance from the wheelchair user during this skill, in the form of having the wheelchair user lean backward or forward at the caregiver's direction, to facilitate the different stages of the skill.

Special considerations for powered wheelchairs operated by users (Version 3)
- The wheelchair user may use the wheelchair's body positioning options (e.g., tilt, recline, leg-rest elevation) to reduce the weight on the smaller wheels.

Special considerations for powered wheelchairs operated by caregivers (Version 4)
- None.

Special considerations for scooters operated by users (Version 5)
- None.

Wheelchair Skills Training
General training tips

- *Adjustment tip*
  - RADs may need to be repositioned or removed to permit caster pops (for manual wheelchairs).
  - RADs may cause the drive wheels to "float" (i.e., with the weight being distributed on the casters and the RADs, unloading the drive wheels).
  - Footrests or anti-tip devices may contact the threshold before the wheels do, making it impossible to negotiate the threshold in that direction without repositioning the wheelchair parts concerned.

- o A seat belt may be useful for higher thresholds to prevent falling out of the wheelchair while the seat is tilted forward.
  - o Wheelchairs with large-diameter leading wheels are able to roll over higher obstacles than those with small-diameter wheels.
  - o Wheelchairs with longer wheelbases are less likely to tip forward as the rear wheels surmount higher thresholds.
- If the wheelchair gets hung up due to insufficient horizontal clearance (wheelbase), the learner may be able to escape by backing up slightly; this will swing the casters from the rear-trailing position to the side- or forward-trailing one, where there is more space between the front and rear wheels.

- *Progression*
  - o The subject should start with low thresholds and progress to higher ones. Obstacles with a height of 10 cm or greater are negotiable in the right wheelchair. Before attempting to negotiate a high obstacle, the subject should be aware of how much clearance exists between the wheels, to avoid getting hung up on the obstacle.

- *Variations*
  - o Leading with the larger-diameter wheels may be helpful.

Special considerations for manual wheelchairs operated by users (Version 1)
- This is the first of a series of skills (including "gets over a gap," "ascends a low curb," and "ascends a high curb") for which the ability to pop the casters in a specific location and move forward are very helpful.

- *Two-hand-propulsion pattern*

  - o *Forward approach, stationary method*
    - – The wheelchair user should approach the obstacle and stop with the casters 5–10 cm from the threshold, to avoid striking the casters on the vertical section of the threshold (Figure 7.79).
    - – This method comprises two phases: "popping" and "leaning". The words "pop" and "lean" can be verbalized as they are performed, as cues.
    - – The wheelchair user first briefly pops the casters from the floor, just high enough to clear the threshold (Figure 7.80). To do so using the two-hand-propulsion method, the wheelchair user applies forward forces of moderate intensity to the hand-rims. After the casters land beyond the threshold, the wheelchair user should lean forward to help power the rear wheels over the threshold (Figure 7.81). Some rocking may be needed.
    - – For higher obstacles, once the rear wheels are on top of the obstacle, the wheelchair user should lean back to decrease the likelihood of a forward tip or fall out of the wheelchair.

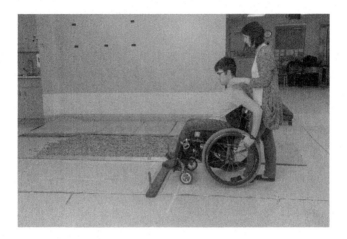

Figure 7.79 Starting position for the stationary approach to the "gets over threshold" skill.

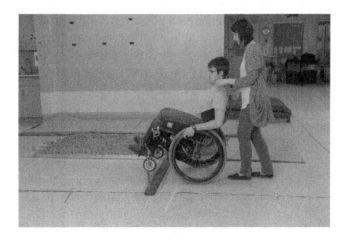

Figure 7.80 Popping the casters over the threshold.

○ *Forward approach, momentum method*
- This method comprises three phases: "coasting," "popping," and "leaning". As for the stationary method, the cues "coast," "pop," and "lean" can be verbalized as they are performed.
- The wheelchair user should initially approach at a slow speed. It is simpler to pop the casters while moving slowly. Also, if the wheelchair user fails to pop the casters for long enough to clear the threshold, the sudden stop will be less jarring at a slow speed.

225

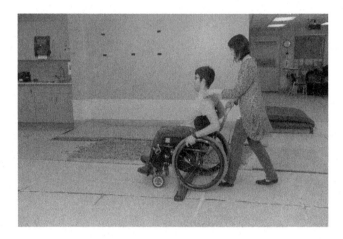

Figure 7.81 Getting the rear wheels over a threshold.

- The wheelchair user should not lean forward to look at the feet as he/she approaches the obstacle, because that increases the weight on the casters. In timing the caster pop, the wheelchair user needs to understand where the casters are (often below the knees, not under the feet). A mirror placed to the side of the obstacle can be used to provide feedback.
- In preparation to pop the front wheels while the wheelchair user moves forward, the wheelchair user briefly coasts to allow correct placement of the hands when he/she is at the proper distance from the threshold. The correct position is when the hands are ready to grasp the hand-rims, behind top dead center (11:00 o'clock on the right wheel, using the clock analogy). Then, the wheelchair user should accelerate the chair even faster than it is coasting, by using a stroke of moderate force that is powerful enough to pop the casters from the surface.
- Once the casters have landed beyond the threshold and the rear wheels strike the threshold, the wheelchair user should lean forward and propel the rear wheels to bring the rear wheels over the obstacle.
- When moving forward over a threshold, some advanced wheelchair users prefer to allow the rear wheels to reach the surface beyond the threshold before having the casters land on the surface. However, when initially learning the skill, it is preferable if the casters land beyond the threshold before the rear wheels strike the obstacle. This will be especially useful in later skills (e.g., ascending curbs) to avoid "caster slap." Caster slap occurs

when the casters are brought forcefully down onto the surface by the deceleration that occurs when the rear wheels strike an obstacle with the casters in the air.

- o *Progression*
  - – To practice getting the timing correct without the fear of having the casters strike the threshold, the wheelchair user may practice propelling the wheelchair forward and transiently popping the casters at a predetermined point on the floor. This can be a line on the floor or a strip of bubble wrap. The horizontal distance over which the casters need to be off the floor can be gradually increased.
  - – The subject should start with the stationary approach then progress to the momentum method.
  - – For learners experiencing difficulties in coordinating the sequence of the three components of the skill (coast, pop, and lean), it may be useful to practice them in segments before putting the segments together.

- o *Variations*
  - – The wheelchair user may find it easier to back over a low obstacle. The wheelchair user should approach the obstacle slowly, because a sudden stop can cause a rear tip. As the wheelchair user approaches the obstacle backward (Figure 7.82), he/she should lean forward to unload the rear wheels and further reduce the likelihood of a rear tip. Using the foot on the floor can give

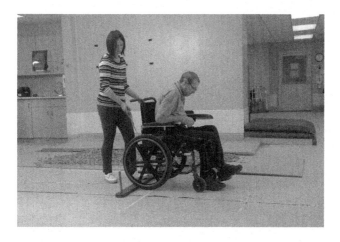

Figure 7.82 Getting over a threshold using the backward approach, leaning forward at the beginning.

the wheelchair user additional power to get over the obstacle (Figure 7.83). The wheelchair user pulls the wheelchair straight backward by applying equal force to both wheels. Otherwise, the casters may turn and catch sideways on the obstacle. Once the rear wheels are over the threshold, the wheelchair user should lean back enough to unload the casters as they reach the obstacle (Figure 7.84), but not so much as to cause a rear tip.

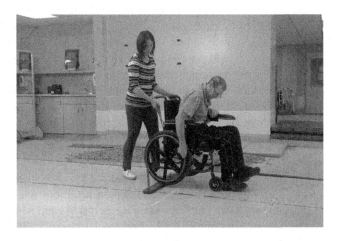

Figure 7.83 Getting over a threshold using the backward approach, with the rear wheels on top.

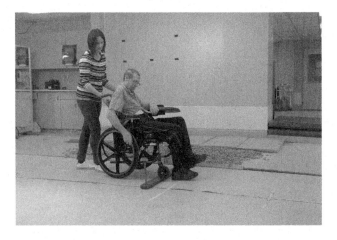

Figure 7.84 Getting over a threshold using the backward approach, leaning backward to unload the casters.

- – The wheelchair user can use a full wheelie for the entire skill or only until the rear wheels strike the threshold.
- – The hands-free version of the skill is useful because the wheels may be spinning too quickly for the hands to catch up with them (e.g., coming down a hill). However, this is an advanced skill. Some wheelchair users can flex the hips, keeping the body upright. Although actively leaning back into the backrest at the intended moment will also pop the casters off the ground, there is an increased risk of the wheelchair user tipping over backward and the body will not be well positioned for the forward lean needed during the second half of this skill. The hands-free version of the skill can be difficult to spot.

- *Hemiplegic-propulsion pattern*
  - o The backward approach (as described above) is useful whenever high-rolling resistance is encountered, as the threshold represents.

  - o *Variations*
    - – The threshold can be approached in the forward direction, using the foot/feet to pop the casters. While popping the casters, at the same time, the wheelchair user should roll the wheelchair forward so that the casters land on the floor beyond the threshold.

Special considerations for manual wheelchairs operated by caregivers (Version 2)
- The caregiver may request assistance from the wheelchair user during this skill, in the form of having the wheelchair user lean backward or forward at the caregiver's direction, to facilitate the different stages of the skill.

Special considerations for powered wheelchairs operated by users (Version 3)
- Positional control (e.g., tilt, recline) can be used to alter the weight distribution of the chair and to provide footrest clearance.
- Smooth continuous forward movement is often the most successful method of traversing a threshold.
- Depending upon the size of the threshold, it may be necessary to switch drive modes to have the necessary wheel torque.
- If the powered wheelchair has come to a stop against the threshold, as extra force is applied to the threshold, the casters may suddenly pop up. The wheelchair user should not apply any more force than is needed and should reduce the force applied to the joystick as soon as possible.
- Getting the larger drive wheels over the threshold is usually easier than getting the smaller caster wheels over. Leaning away from the casters will unload them and make it easier to get them over.

Special considerations for powered wheelchairs operated by caregivers (Version 4)
- None.

Special considerations for scooters operated by users (Version 5)
- If there is insufficient vertical ground clearance between the front and rear wheels, the scooter may get hung up on a high threshold.
- Approaching the threshold with a little extra speed may help. However, if the scooter user approaches the threshold too quickly, the stiffness of the suspension may cause the scooter user to bounce off the seat and lose control of the scooter.

7.28 Gets over gap

Versions applicable

Version	Wheelchair Type	Subject Type	Applicable
1	Manual	Wheelchair user	✓
2		Caregiver	✓
3	Powered	Wheelchair user	✓
4		Caregiver	✓
5	Scooter	Wheelchair user	✓

Skill level
- Intermediate.

Description
- The subject moves the wheelchair over a gap across the line of progression.

Rationale
- A gap in surface support is a commonly encountered barrier (e.g., due to a rut in the road, a water channel, a space between a subway platform and the train). Small gaps, that only affect one wheel at a time, may be jarring but are not usually major obstacles. In this section, only gaps that are as wide as the wheelchair will be considered. Small-diameter wheels such as casters can drop into such gaps, causing a sudden deceleration that can tip the wheelchair over forward or lead to the wheelchair user falling out of the wheelchair. Even if no tip or fall occurs, it can be difficult to get the wheelchair out of the gap.

Prerequisites
- None.

Spotter considerations
- Spotter starting position: If using a single spotter, he/she should be behind the wheelchair, holding onto a spotter strap with one hand (if a manual wheelchair) and with the other hand in front of the wheelchair user's shoulder. If using two spotters (as is recommended), the second spotter should stand to one side of the gap. A removable seat belt can prevent the wheelchair user from falling from the wheelchair.
- Risks requiring spotter intervention:
  - Rear tip when accelerating to pop the casters from the surface (if a manual wheelchair).
  - Forward tip or fall if the casters roll or drop into the gap.

Wheelchair Skills Test (WST)

Equipment
- Smooth level surface 1.5 m wide, with at least 1.5 m before and after the gap.
- The gap should be approximately 5 cm deep, the full width of the path and 15 cm across (in the line of progression).
- If a gap is not readily available, one can be easily simulated. For instance, two folding tables (with the legs folded) or two gym mats can be put close together.

Starting positions
- Wheelchair: Leading wheels at least 0.5 m in front of the gap.

Instructions to subject
- *"Get your wheelchair over the gap* (indicate it)."

Capacity criteria
- As for the general scoring criteria, with the clarifications below.
- A "pass" score should be awarded if all components of the wheelchair are on the level surface beyond the gap.
- A "pass with difficulty" score should be awarded if:
  - There is significant jarring.
  - There is any loss of control due to bouncing off the seat.
  - There is unintended hyper-flexion of the lower limb without injury.

Special considerations for manual wheelchairs operated by users (Version 1)
- The wheelchair user is permitted to use his/her feet or stand to get over the gap.

Special considerations for manual wheelchairs operated by caregivers (Version 2)
- As for "gets over threshold."

Special considerations for powered wheelchairs operated by users (Version 3)
- The wheelchair user may use the wheelchair's body positioning options (e.g., tilt, recline, leg-rest elevation) to reduce the weight on the smaller wheels.

Special considerations for powered wheelchairs operated by caregivers (Version 4)
- None.

Special considerations for scooters operated by users (Version 5)
- None.

Wheelchair Skills Training
General training tips

- *Adjustment tip*
  - The diameter of the wheels affects the size and depth of gaps that can be overcome.
- The best approach is to avoid gaps, steering around them, or straddling them.
- The wheelchair user may approach the gap squarely or obliquely.
- If the casters drop into the gap and they turn sideways (a common problem if the wheelchair is moved forward and backward repeatedly in an attempt to get the casters out of the gap), it can be very difficult or impossible to proceed without assistance.

- *Progression*
  - The subject should start with a slow speed and increase speed.
  - The subject should start with small shallow gaps and progress to more challenging ones.

- *Variations*
  - As long as three wheels are supported at any time, the wheelchair will usually remain upright. That being the case, an oblique approach to a gap (e.g., 30°–45° from the line of progression so that only one wheel is unsupported at a time) is a useful strategy. The wheelchair user should keep his/her weight away from the unsupported wheel.

Special considerations for manual wheelchairs operated by users (Version 1)
- This skill builds on the caster popping practice done in the "rolls on soft surface" skill earlier.
- See the "gets over a threshold" skill.
- The square approach (Figure 7.85a through d) is useful to include in training because the method used is part of a step-wise sequence leading toward the ascent of level changes and curbs.

- *Variations*
  - When popping the casters over a long gap, the wheelchair user can use the full wheelie position or perform a transient pop with two pushes, the second push while the casters are still in the air.

Special considerations for manual wheelchairs operated by caregivers (Version 2)
- The caregiver may proceed in the forward direction, using the transient caster pop or full wheelie method.
- After the rear wheels are in the gap, the casters can be lowered to the surface beyond the gap. Then, the wheelchair user is asked to lean forward and the wheelchair is rolled out of the gap.

(a)

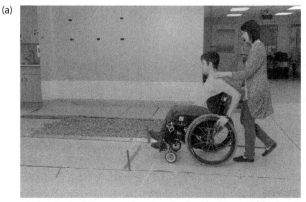

(b)

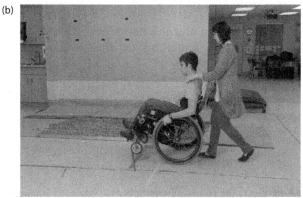

Figure 7.85 "Gets over gap" skill in manual wheelchair: (a) coasting and (b) popping casters over the gap. *(Continued)*

- *Variations*
  - The backward direction may be easier for the caregiver. If this technique is used, the rear wheels of the wheelchair can be lowered into the pothole, then the wheelchair tipped into a wheelie position to be pulled out of the pothole on the rear wheels.

Special considerations for powered wheelchairs operated by users (Version 3)

- *Adjustment tip*
  - If the casters are rounded on their sides (i.e., ball shaped), they will better resist the tendency to get caught sideways.
- Positional control (e.g., tilt, recline) can be used to get the weight over the drive wheels and improve traction.

234

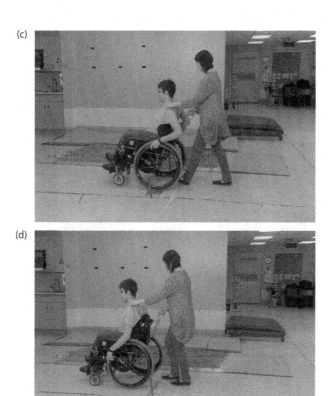

(c)

(d)

Figure 7.85 (Continued) (c) Rear wheels in gap and (d) rear wheels over gap.

- If the gap cannot be managed in the oblique direction or avoided but appears to be negotiable straightforward, it is best to proceed at a slow speed but a steady pace because the momentum may help to bounce the wheels over the gap (Figure 7.86a through d).

Special considerations for powered wheelchairs operated by caregivers (Version 4)

- If the casters get stuck sideways in the gap (Figure 7.87), the caregiver may need to stand on the back of the wheelchair to tilt the chair enough to get the casters out of the gap (Figure 7.88). If the wheelchair user cannot operate the joystick enough to help, a second caregiver may be needed. The motors may need to be disengaged to allow the wheelchair to be pushed out of the gap.

Special considerations for scooters operated by users (Version 5)

- None.

235

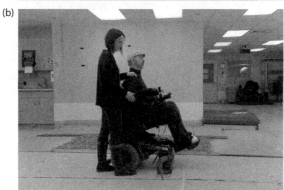

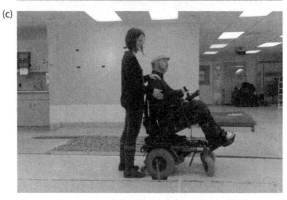

Figure 7.86 "Gets over gap" skill in powered wheelchair: (a) starting position, (b) casters in the gap, and (c) rear wheels in the gap. (*Continued*)

Figure 7.86 (Continued) (d) Rear wheels over the gap.

Figure 7.87 Casters stuck in gap sideways.

Figure 7.88 Caregiver standing on back of powered wheelchair to unload casters in gap.

7.29 Ascends low curb

Versions applicable

Version	Wheelchair Type	Subject Type	Applicable
1	Manual	Wheelchair user	✓
2		Caregiver	✓
3	Powered	Wheelchair user	✓
4		Caregiver	✓
5	Scooter	Wheelchair user	✓

Skill level
- Intermediate.

Description
- The subject gets the wheelchair up a low curb.

Rationale
- Level changes (e.g., curbs, steps, home entries, uneven sidewalk sections) are common obstacles in the natural and built environments.

Prerequisites
- None.

Spotter considerations
- Spotter starting position: For this and later curb-handling and stairs skills, the spotter strap is of little use if a sideways tip or fall occurs. If using a single spotter, he/she should be behind the wheelchair, with both hands close to the push-handles (if any). If using two spotters (as is recommended), the second spotter should stand to one side of the level change. A removable seat belt can prevent the wheelchair user from falling from the wheelchair.
- Risks requiring spotter intervention:
  - Rear tip when accelerating to pop casters from surface (if a manual wheelchair).
  - Forward tip or fall if casters strike the curb.
  - Sideways tip if one wheel gets up onto the upper level before the other.

Wheelchair Skills Test (WST)

Equipment
- The pathway on the lower level leading to the curb should be at least 1.5 m wide and at least 3 m long, for subjects who use a moving approach. The pathway on the upper level leading from the curb edge should be at least 1.5 m wide and at least 1.5 m long.

- The curb should be 5 cm high.
- The nosing of the curb should be gently rounded.
- Bracing or weighting may be needed to prevent the curb from moving when struck by the wheelchair.

Starting positions
- Wheelchair: All wheels are on the level surface below the curb, facing the curb and at least 0.5 m from it. If the subject uses a moving approach, the subject may choose to begin farther away.

Instructions to subject
- *"Get the wheelchair up on the curb."*

Capacity criteria
- As for the general scoring criteria, with the clarifications below
- A "pass" score should be awarded if:
  - All wheels are on the top surface, with the wheelchair user seated upright in the wheelchair.
  - The subject may remove the footrests and reposition the RADs but must be able to do so independently.
  - The wheelchair user may get out of the wheelchair to accomplish the task, if he/she can do so safely.
  - Curb-climbing aids may be used if the wheelchair is equipped with these devices, but the subject must be able to activate and inactivate the aids independently.
- A "pass with difficulty" should be awarded if:
  - There is significant jarring.
  - There is any loss of control due to bouncing off the seat.
  - There is unintended hyper-flexion of the lower limb without injury.

Special considerations for manual wheelchairs operated by users (Version 1)
- A "pass with difficulty" should be awarded if there is a minor thumb injury due to contact with the wheel locks.

Special considerations for manual wheelchairs operated by caregivers (Version 2)
- The caregiver may request assistance from the wheelchair user during this skill, in the form of having the wheelchair user lean backward or forward at the caregiver's direction, to facilitate the different stages of the skill.
- A "pass with difficulty" should be awarded if a caregiver uses poor ergonomic technique (e.g., lifting rather than rolling the wheelchair up onto the upper level).

Special considerations for powered wheelchairs operated by users (Version 3)
- The wheelchair user may use the wheelchair's body positioning options (e.g., tilt, recline, leg-rest elevation) to reduce the weight on the smaller wheels.

Special considerations for powered wheelchairs operated by caregivers (Version 4)
- None.

Special considerations for scooters operated by users (Version 5)
- None.

Wheelchair Skills Training

General training tips
- As for the "gets over threshold" skill, the footrests, anti-tip devices, and clearance between the wheels may affect the ability to negotiate level changes.
- It may be necessary to reposition or remove the footrests or RADs.
- This skill is similar to and builds on the previous ones, specifically the soft surface, threshold, and gap skills.

- *Progression*
  - The subject should start with a minimal level change and progress to higher ones.

Special considerations for manual wheelchairs operated by users (Version 1)

- *Two-hand-propulsion pattern*
  - This skill is similar to the "gets over threshold" and "gets over gap" skills in that it can be approached with stationary and momentum methods.
  - It is slightly more challenging to deal with the rear wheels than the preceding skills because the tilted position due to having the casters on top of the curb moves more weight to the back of the wheelchair (Figure 7.89a through e). This shift of weight is present until the rear wheels are all the way up on the upper level.
  - In the stationary approach, if the wheelchair user has difficulty getting the rear wheels up onto the upper level, the wheelchair user should roll the wheelchair backward until the front wheels are almost off the edge of the curb. This has two effects. First, it reverses the caster trail, thereby reducing the extent of rear tip (because the caster stems are no longer vertical). This provides a greater safety margin between the resting position and the rear tip-over threshold, so the wheelchair user can push harder without tipping over. Second, because the rear

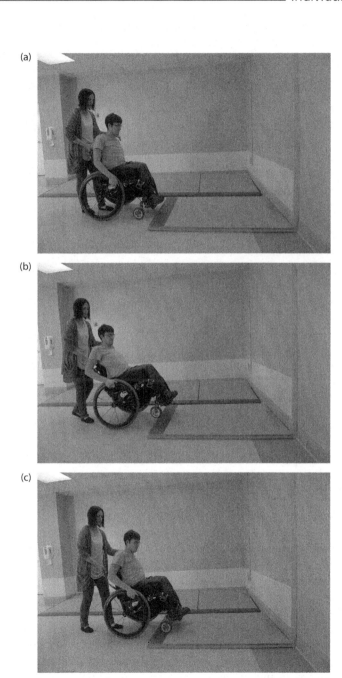

Figure 7.89 For the "ascends low curb" skill in the forward direction: (a) starting position for the stationary approach, (b) popping casters, and (c) landing casters on top of curb. (*Continued*)

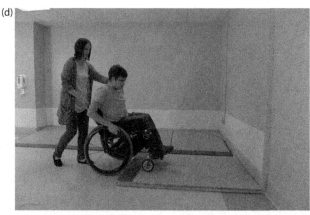

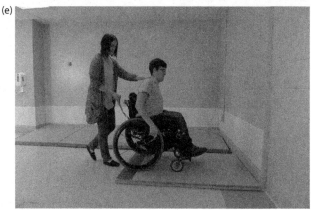

Figure 7.89 (Continued) (d) Rear wheels against curb edge and (e) rear wheels up curb.

wheels have been backed slightly away from the edge, a small amount of momentum can be used. Before backing the rear wheels away from the curb edge, the wheelchair user should place his/her hands on the hand-rims in the position where the most force can be applied. The hands should remain on the hand-rims as the rear wheels are backed away from the curb, ensuring that the hands and trunk will be optimally placed when moving forward again. When the rear wheels strike the curb, the wheelchair user should lean forward and push the rear wheels up onto the upper level. The forward lean should be timed to coincide with when the rear wheels contact the curb.

o As noted earlier, with the momentum method, the wheelchair user should ensure the casters are on the upper surface (rather than in the air) before the rear wheels hit the lip of the curb. If the casters are

still in the air, the energy from the forward pitch caused by the collision of the rear wheels with the obstacle will be expended in noisily
bringing the casters down on the upper level ("caster slap") rather
than bringing the rear wheels up onto the upper level.

- o *Variations*
  - – The wheelchair user might find it easier to ascend the 5 cm curb
    backward (see below).
  - – The wheelchair user may use the external environment if avail-
    able (e.g., door frame, street pole).
- *Hemiplegic-propulsion pattern*
  - o The wheelchair is backed up until the rear wheels contact the obstacle.
    Then, leaning forward to unload the rear wheels, the foot on the floor is
    used to push the rear wheels up the level change. Then, the wheelchair
    user sits upright and pushes down on the foot on the floor or top of the
    curb to bring the casters up to the upper level (Figure 7.90a through c).

**Special considerations for manual wheelchairs operated by caregivers
(Version 2)**
- To ascend a level change forward, as shown in Figure 7.96 for the "ascends
  high curb" skill, the caregiver should put the wheelchair into the full or
  partial wheelie position to get the casters onto the upper level. Then, the
  caregiver should roll the chair forward until the rear wheels touch the vertical edge of the level change. Then, the caregiver should ask the wheelchair
  user to lean forward to reduce the weight on the rear wheels. The caregiver
  then applies a forward and upward force on the push-handles or some other
  rigid part of the wheelchair to help the rear wheels roll up onto the upper
  level. Once on the upper level, the wheelchair user may sit upright again.

- *Variations*
  - o For a small level change, the caregiver can bring the wheelchair up the
    curb backward. If the level change is large enough, the caregiver may
    need to tip the wheelchair into the full wheelie position (to avoid tipping
    the wheelchair user forward out of the wheelchair) and pull the wheel-
    chair up onto the upper level. The caregiver should step well away from
    the edge of the level change before lowering the casters. The caregiver
    should not use this technique for a large level change, because he/she
    would need to bend forward too far and might injure his/her back.

**Special considerations for powered wheelchairs operated by users
(Version 3)**
- Positional control (e.g., tilt, recline) can be used to alter the weight distri-
  bution on the wheels and to provide footrest clearance.
- Smooth continuous forward movement is often the most successful
  method of ascending a curb (Figure 7.91a through c).

243

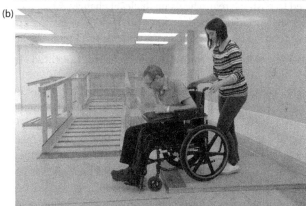

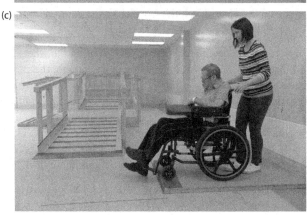

Figure 7.90 For the "ascends low curb" skill in the backward direction: (a) rear wheels against curb, leaning forward, (b) rear wheels on top of curb, and (c) leaning back to get the casters up the curb.

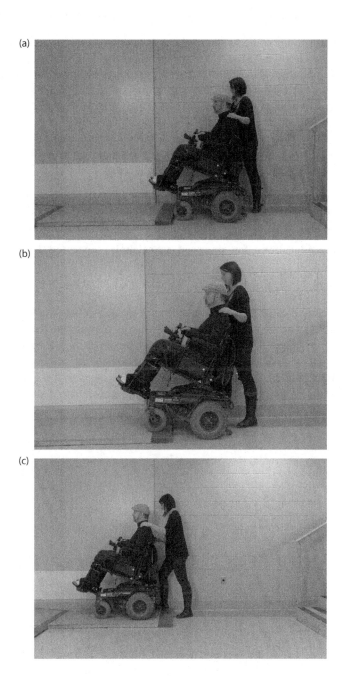

Figure 7.91 For the "ascends low curb" skill in the forward direction in a powered wheelchair: (a) approach squarely, (b) casters on top of curb, and (c) rear wheels on top of curb.

- Depending upon the height of the curb, it may be necessary to switch drive modes to have the necessary wheel torque.
- If the powered wheelchair has come to a stop against the curb, as extra force is applied to the curb, the casters may suddenly pop up. The wheelchair user should not apply any more force than is needed and should reduce the force applied to the joystick as soon as possible.
- Getting the larger drive wheels up the curb is usually easier than getting the smaller caster wheels up. Leaning away from the casters will unload them and make it easier to get them up the curb.

- *Variation*
  - In some instances, especially with a rear-wheel-drive wheelchair, it may be easier to ascend the level change in the reverse direction.

Special considerations for powered wheelchairs operated by caregivers (Version 4)
- None.

Special considerations for scooters operated by users (Version 5)
- If there is insufficient vertical ground clearance between the front and rear wheels (Figure 7.92), the scooter may get hung up on the edge of the curb.
- Approaching the low curb with a little extra speed may help to mount the curb. However, if the scooter user approaches the curb too quickly, the stiffness of the suspension may cause the scooter user to bounce off the seat and lose control of the scooter.

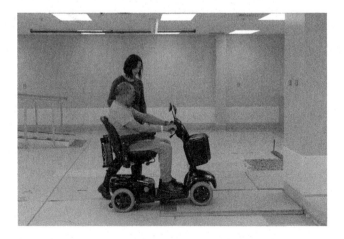

Figure 7.92 Scooter partway up low curb.

7.30 Descends low curb

Versions applicable

Version	Wheelchair Type	Subject Type	Applicable
1	Manual	Wheelchair user	✓
2		Caregiver	✓
3	Powered	Wheelchair user	✓
4		Caregiver	✓
5	Scooter	Wheelchair user	✓

Skill level
- Intermediate.

Description
- The subject gets the wheelchair down a low curb.

Rationale
- As for the "ascends low curb" skill.

Prerequisites
- None.

Spotter considerations
- Spotter starting position:
  - If the wheelchair user uses the forward direction, simply rolling off the low curb, the spotter may simply stand on the lower level close enough to intervene if the wheelchair tips forward or the wheelchair user falls from the wheelchair. A removable seat belt can prevent the latter.
  - If the task is performed forward in the wheelie position, the spotter should be behind the wheelchair, with one hand close to each push-handle (if a manual wheelchair). If a second spotter is available, he/she should be on the lower level or a seat belt should be used.
  - If the wheelchair user uses the backward technique, the spotter should be standing on the lower level with the hands positioned near the push-handles (if a manual wheelchair).
- Risks requiring spotter intervention:
  - Rear tip if performed backward or forward in the wheelie position.
  - Forward tip or fall from the wheelchair if the task is performed by rolling forward off the curb.
  - Sideways tip if one wheel drops off the upper level before the other.

Wheelchair Skills Test (WST)

Equipment
- As for "ascends low curb" skill except, because many subjects can descend level changes from a higher level than they can ascend, some

alternative means (e.g., an incline) of getting to the upper level is recommended. The tester can help to get the wheelchair to the upper level.

Starting positions
- Wheelchair: All wheels are on the level surface above the curb edge, facing the edge, with the leading wheels at least 0.5 m away from it.

Instructions to subject
- *"Get the wheelchair down to the lower level."*

Capacity criteria
- As for the general scoring criteria, with the clarifications below.
- A "pass" should be awarded if:
  - All wheels are on the lower level, the wheelchair user is seated upright in the wheelchair and the wheelchair is free to roll away (i.e., not hung up on the footrests or RADs).
  - Any technique is permitted.
  - The wheelchair user may get out of the wheelchair to accomplish the task, if he/she can do so safely.
  - The subject may remove the footrests and reposition the RADs but must be able to do so independently.

Special considerations for manual wheelchairs operated by users (Version 1)
- None.

Special considerations for manual wheelchairs operated by caregivers (Version 2)
- The caregiver may request assistance from the wheelchair user during this skill, in the form of having the wheelchair user lean backward or forward at the caregiver's direction, to facilitate the different stages of the skill.

Special considerations for powered wheelchairs operated by users (Version 3)
- None.

Special considerations for powered wheelchairs operated by caregivers (Version 4)
- None.

Special considerations for scooters operated by users (Version 5)
- None.

Wheelchair Skills Training

General training tips
- For a low curb, forward or backward are both appropriate approaches.

248

- The wheelchair may be able to simply roll forward off the upper level. This is less of a problem for wheelchairs with long wheelbases.
- It may be as safe and effective to go off the lip at a moderate or full speed as it is to go slowly.

Special considerations for manual wheelchairs operated by users (Version 1)

- *Two-hand-propulsion pattern*
  - The forward approach is convenient and allows the subject to watch for traffic (Figure 7.93a through c).

  - *Variations*
    - If the footrests catch on the ground or there is the danger of a forward tip or fall from the wheelchair, the wheelchair user may use the backward approach (Figure 7.94a through c). Learning the backward approach will be helpful when advancing to higher curbs. The wheelchair user should line the rear wheels up with the edge of the curb. The wheelchair user should lean as far forward as possible (chest on lap), and reach forward on the hand-rims. The wheelchair user should move backward very slowly and let the rear wheels roll evenly down off the upper level under control. Once the rear wheels are on the lower level, the wheelchair user can sit more upright if this is possible without tipping over back-ward. The wheelchair user should avoid braking suddenly when the rear wheels land on the lower level because this can induce a rear tip; keeping the wheelchair moving backward reduces the likelihood of this problem. If the wheelchair can be brought to a stop with the rear wheels on the lower level and the casters on the upper level, the wheelchair user can turn to the left or the right to get the casters off the upper level without scraping the footrests— by the time the second caster rolls off the edge, the footrests are beyond the edge. Alternatively, the wheelchair user can use the full wheelie position to move backward away from the curb.
    - Approaching the curb edge in the forward direction, the wheelchair user can transiently pop the casters as they reach the curb edge.
    - The wheelchair user can use the full wheelie position. This is discussed in more detail later in the section on the "descends high curb in wheelie position" skill.

- *Hemiplegic-propulsion pattern*
  - The wheelchair is moved forward to the edge of the curb. Then, lean-ing backward to avoid a forward tip or falling out of the wheelchair, the foot is placed on the surface below the curb. The wheelchair is moved slowly forward until the rear wheels are on the surface below the curb (Figure 7.95a through c).

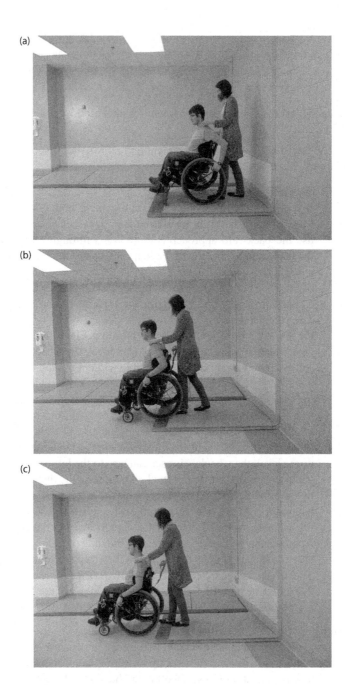

Figure 7.93 For the "descends low curb" skill, in the forward direction: (a) starting position, (b) casters on lower level, and (c) rear wheels on lower level.

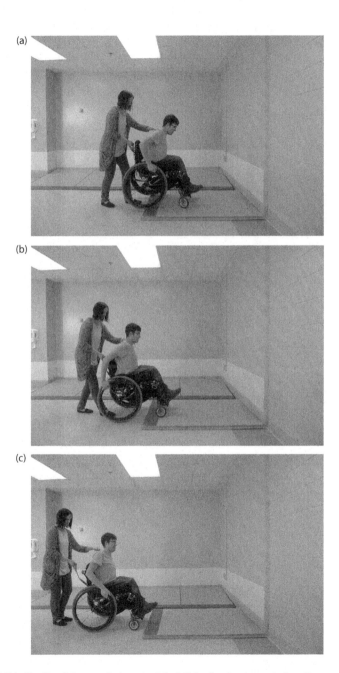

Figure 7.94 For the "descends low curb" skill in the backward direction:
(a) leaning forward as rear wheels approach the edge, (b) sitting more upright
when the rear wheels are on the lower level, and (c) casters on the lower level.

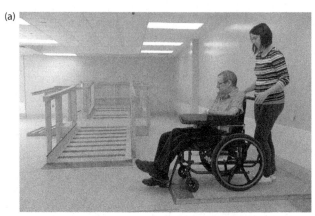

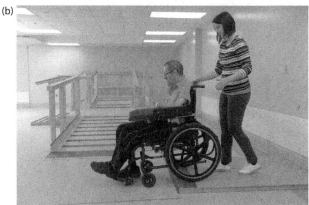

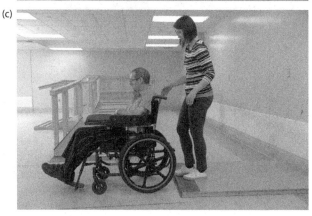

Figure 7.95 For the "descends low curb" skill in the forward direction by a person with hemiplegia: (a) starting position, (b) casters on the lower level, and (c) rear wheels on the lower level.

Special considerations for manual wheelchairs operated by caregivers (Version 2)

- In the forward direction, the caregiver may slowly push the wheelchair off the upper level, allowing the casters to gently land on the lower level, followed by the rear wheels. It is dangerous for the caregiver to use this technique for medium or large level changes—the wheelchair user may tip forward out of the wheelchair or the footrests may dig in and prevent a smooth descent.

- *Variations*
  - To descend a level change backward, as shown in Figure 7.97 for the "descends high curb" skill, the caregiver should turn the wheelchair around so that the rear wheels go off the edge first. The caregiver should stand close behind the wheelchair and on the lower level. The caregiver should align the rear wheels so that they are both on the edge of the upper level. The caregiver then asks the wheelchair user to lean forward to reduce the weight on the rear wheels. Controlling the movement of the chair, the caregiver should slowly and evenly roll the rear wheels down onto the lower level, avoiding any jarring. Leaning the caregiver's torso against the backrest is acceptable. Once the rear wheels are on the lower level and the wheelchair user is sitting upright, the caregiver may need to tip the wheelchair back into the wheelie position to avoid the footrests scraping on the upper level during the caster descent. Alternatively, the caregiver can turn the chair sideways to prevent the footrests from getting caught.
  - Approaching the curb edge in the forward direction, the caregiver can tip the wheelchair back into the full wheelie position and lower the rear wheels to the lower level. The caregiver should be careful about the extent to which his/her back is flexed. However, this technique has the advantage of allowing continuous progression along a street, with the caregiver's eyes facing any dangers in traffic. The caregiver should not attempt to descend the level change backward with the wheelchair in the wheelie position because, at greater heights, this causes severe jarring of the wheelchair and its occupant.

Special considerations for powered wheelchairs operated by users (Version 3)

- None.

Special considerations for powered wheelchairs operated by caregivers (Version 4)

- None.

Special considerations for scooters operated by users (Version 5)

- If there is insufficient vertical ground clearance between the front and rear wheels, the scooter may get hung up on the edge of the curb.

7.31 Ascends high curb

Versions applicable

Version	Wheelchair Type	Subject Type	Applicable
1	Manual	Wheelchair user	✓
2		Caregiver	✓
3	Powered	Wheelchair user	
4		Caregiver	
5	Scooter	Wheelchair user	

Skill level
- Advanced.

Description
- The subject ascends a high curb.

Rationale
- As for the "ascends low curb" skill. Although curb cuts ("pedestrian ramps") are now commonplace in many parts of the world, curbs or large level changes are still commonly encountered. This skill is not applicable for most powered wheelchairs and scooters because of the difficulty and danger involved.

Prerequisites
- "Ascends low curb" skill.

Spotter considerations
- As for "ascends low curb" skill.

Wheelchair Skills Test (WST)

Equipment
- As for "ascends low curb" skill except 15 cm high.

Starting positions
- As for "ascends low curb" skill.

Instructions to subject
- *"Get the wheelchair up on the curb."*

Capacity criteria
- As for the general scoring criteria, with the clarifications below.
- As for "ascends low curb" skill.
- A "fail" score should be awarded if the subject has failed the "ascends low curb" skill.

Special considerations for manual wheelchairs operated by users (Version 1)
- None.

Special considerations for manual wheelchairs operated by caregivers (Version 2)
- As for "ascends low curb" skill.
- A "pass with difficulty" score should be awarded if:
  - The caregiver fails to have the wheelchair user lean forward while rolling the rear wheels forward up the curb.
  - The caregiver lifts rather than rolls the wheelchair to the upper level.
  - The caregiver pulls the wheelchair up the curb backward in the wheelie position, which is ergonomically unsound due to the amount of forward trunk flexion and high forces required unless the wheelchair user is light and the caregiver is strong.

Special considerations for powered wheelchairs operated by users (Version 3)
- Not applicable.

Special considerations for powered wheelchairs operated by caregivers (Version 4)
- Not applicable.

Special considerations for scooters operated by users (Version 5)
- Not applicable.

Wheelchair Skills Training

General training tips

- As for the "ascends low curb" skill.

- *Progression*
  - It is useful to have a 10 cm curb as an intermediate height between the low (5 cm) and high (15 cm) curbs.
  - The subject should begin with the stationary method and a small curb height.
  - The subject should gradually increase the height of the curb until it becomes difficult to get the rear wheels up on top of the curb.
  - The subject should then change to the momentum approach with a small curb height and gradually increase the height.
  - The subject should reduce the distance available for the approach.

    ○ For subjects and wheelchairs capable of handling curbs higher than 15 cm, it is reasonable to attempt these under the supervision of a trainer, if it can be done safely.

- *Variations*
  - ○ The subject should perform the skill in a setting that includes an element of side slope on the approach. This requires the wheelchair user to anticipate the amount of downhill-turning tendency that will occur during the coast phase of the skill.

Special considerations for manual wheelchairs operated by users (Version 1)
- As for the "ascends low curb" skill, for both two-hand-propulsion and hemiplegic-propulsion patterns.

Special considerations for manual wheelchairs operated by caregivers (Version 2)
- As for the "ascends low curb" skill (Figure 7.96a through f).

Special considerations for powered wheelchairs operated by users (Version 3)
- Not applicable.

Special considerations for powered wheelchairs operated by caregivers (Version 4)
- Not applicable.

Special considerations for scooters operated by users (Version 5)
- Not applicable.

(a)

Figure 7.96 For the "ascends high curb" skill in the forward direction by a caregiver: (a) foot on tipping lever. *(Continued)*

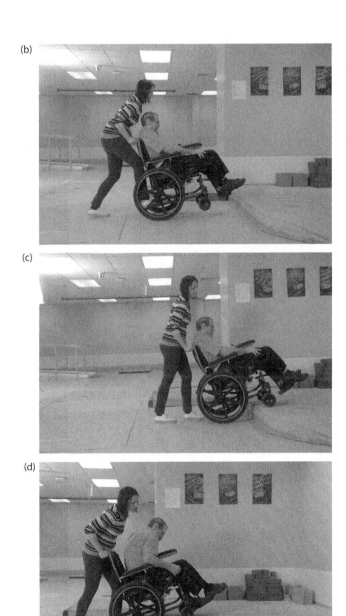

Figure 7.96 (Continued) (b) Tilting the wheelchair, (c) casters on the upper level, and (d) rear wheels against the curb edge with the wheelchair user leaning forward. (Continued)

(e)

(f)

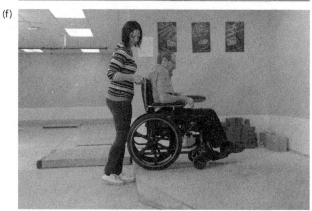

Figure 7.96 (Continued) (e) Rolling the rear wheels up the curb and (f) finish position.

7.32 Descends high curb

Versions applicable

Version	Wheelchair Type	Subject Type	Applicable
1	Manual	Wheelchair user	✓
2		Caregiver	✓
3	Powered	Wheelchair user	
4		Caregiver	
5	Scooter	Wheelchair user	

Skill level
- Advanced.

Description
- The subject gets the wheelchair down a high curb.

Rationale
- As for the "descends a low curb" skill. The appropriate technique for a high curb differs in some respects from that used for a lower curb height. This skill is not applicable for most powered wheelchairs and scooters because of the difficulty and danger involved.

Prerequisites
- "Descends low curb" skill.

Spotter considerations
- Spotter starting position:
  - Behind the wheelchair, with the hands near the push-handles of the wheelchair.
  - For the forward-wheelie approach, if using two spotters, the second spotter should stand beside and below the curb. A removable seat belt may be helpful.
- Risks requiring spotter intervention:
  - Rear tip if performed backward or in the forward direction in the wheelie position.
  - Forward tip or fall if performed by rolling forward off the level change (not generally recommended unless the wheelbase is long).
  - Sideways tip if one wheel drops off the upper level before the other.
  - Serious jarring if a caregiver attempts to bring the wheelchair off the curb backward in the wheelie position.

Wheelchair Skills Test (WST)

Equipment
- As for "ascends high curb" skill.

Starting positions
- Wheelchair: The leading wheels at least 0.5 m from the curb edge.

Instructions to subject
- Screening questions ("*Can you do it? How?*") are strongly recommended before the subject is allowed to proceed to the objective testing of this skill. If a method is described that may not be unsafe but that the tester has concerns about from the perspective of being able to spot the skill in a manner that is safe for both the subject and personnel, the tester may allow the subject to choose another method without penalty.
- "*Get the wheelchair down the curb.*"

Capacity criteria
- As for the general scoring criteria, with the clarifications below.
- Except as noted below, as for the "descends low curb" skill.
- A "fail" score should be awarded if:
  - On the screening questions, the subject is unable to describe an acceptable method of performing the skill.
  - The subject fails the "descends low curb" skill.
  - The subject is about to allow one wheel to drop off the upper level before the other. The tester or trainer should intervene to prevent completion of such an attempt.

Special considerations for manual wheelchairs operated by users (Version 1)
- None.

Special considerations for manual wheelchairs operated by caregivers (Version 2)
- Except as noted below, as for the "descends low curb" skill.
- It is permissible for a caregiver to use the wheelie position to lower the wheelchair in the forward direction.
- A "pass with difficulty" score should be awarded if:
  - There is significant jarring due to an uncontrolled drop of the wheels to lower level.
  - The caregiver fails to have the wheelchair user lean forward while rolling the rear wheels backward down the curb.
- A "fail" score should be awarded if a caregiver attempts to bring the wheelchair off the curb backward in the wheelie position. The tester should intervene.

Special considerations for powered wheelchairs operated by users (Version 3)
- Not applicable.

Special considerations for powered wheelchairs operated by caregivers (Version 4)
- Not applicable.

Special considerations for scooters operated by users (Version 5)
- Not applicable.

Wheelchair Skills Training

General training tips
- As for the "descends low curb" skill.

Special considerations for manual wheelchairs operated by users (Version 1)

- *Two-hand-propulsion pattern*
    - o The backward approach (see the "descends low curb" skill) is simple and generally safe if the wheelchair has adequate rear stability. For this skill, it is especially important to practice with a spotter until it has been mastered.

    - o *Variations*
        - – The forward curb descent in the wheelie position is dealt with later in the "descends high curb in wheelie position" skill section.
        - – The forward, transient-wheelie method is an advanced skill especially from this height. As for this variation described earlier for the "descends low curb" skill, the wheelchair user approaches the curb edge squarely with all four wheels on the surface and pops the casters as they reach the edge. This is similar to the technique used to pop the casters for the "gets over threshold" and "gets over gap" skills. The extent of the caster pop should be sufficient to allow the rear wheels to land on the lower level at about the same time or slightly before the casters land. This method requires good timing and skill, but is a natural way to maintain forward progression and to watch for traffic. It can be difficult to spot, so two spotters are recommended.

- *Hemiplegic-propulsion pattern*
    - o As for the "descends low curb" skill.

Special considerations for manual wheelchairs operated by caregivers (Version 2)
- As for the "descends low curb" skill (Figure 7.97a through d).

Special considerations for powered wheelchairs operated by users (Version 3)
- Not applicable.

Special considerations for powered wheelchairs operated by caregivers (Version 4)
- Not applicable.

Special considerations for scooters operated by users (Version 5)
- Not applicable.

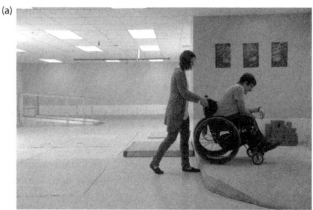

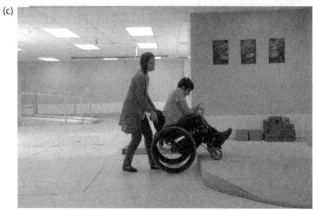

Figure 7.97 For the "descends high curb" skill in the backward direction by a caregiver: (a) starting position, (b) lowering the rear wheels to the lower level, and (c) Tilting the wheelchair back. *(Continued)*

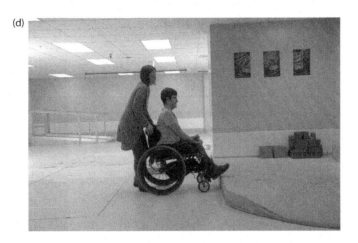

Figure 7.97 (Continued) (d) lowering the casters.

7.33 Performs stationary wheelie

Versions applicable

Version	Wheelchair Type	Subject Type	Applicable
1	Manual	Wheelchair user	✓
2		Caregiver	✓
3	Powered	Wheelchair user	
4		Caregiver	
5	Scooter	Wheelchair user	

Skill level
- Advanced.

Description
- The subject achieves the wheelie position (balancing on the rear wheels), maintains it for a period of time and brings the casters back to the floor.

Rationale
- The stationary wheelie is a foundation skill for a number of functional skills that can be best performed in the full wheelie position, skills such as descent of a steep incline or descent of a high curb. The stationary wheelie position can also be used to avoid postural problems that can cause neck strain from looking up. This skill is not applicable for most powered wheelchairs and scooters.

Prerequisites
- None.

Spotter considerations
- Spotter starting position: Usually the spotter stands behind the wheelchair holding onto a spotter strap. However, the skill can also be spotted from a position in front and to one side of the wheelchair (Figure 6.6), with a hand ready to apply a downward and backward force to the wheelchair user's knee or a fixed part of the wheelchair.
- Risks requiring spotter intervention: Rear tip if the subject overshoots on takeoff or loses balance.

Wheelchair Skills Test (WST)

Equipment
- As for the "turns in place" skill.

Starting positions
- Wheelchair: In the center of the square.

Instructions to subject
- *"Get the wheelchair into the wheelie position and hold it until I tell you to stop. Keep the rear wheels within the box* (indicate it)."
- After 30 seconds, *"Come down now."*

Capacity criteria
- As for the general scoring criteria, with the clarifications below.
- A "pass" should be awarded if:
  - o The subject achieves the wheelie position and holds this position in a controlled manner for 30 seconds while all wheelchair parts that are in contact with the floor remain within the square.
  - o After 30 seconds, a controlled return to the upright position is made. The subject must wait for the instruction to bring the casters back to the floor before doing so. The casters must land inside the square.
  - o It is permissible to use the feet to achieve the wheelie position but not to maintain it.
- A "pass with difficulty" should be awarded if:
  - o An aided wheelie is used (casters off the floor, balanced on RADs).
  - o There is significant jarring because the subject lands too vigorously.

Special considerations for manual wheelchairs operated by users (Version 1)
- None.

Special considerations for manual wheelchairs operated by caregivers (Version 2)
- There is no need for the caregiver to maintain the wheelie for 30 seconds as long as the tester is satisfied that the caregiver has achieved the balance position correctly and is capable of maintaining it.

Special considerations for powered wheelchairs operated by users (Version 3)
- Not applicable.

Special considerations for powered wheelchairs operated by caregivers (Version 4)
- Not applicable.

Special considerations for scooters operated by users (Version 5)
- Not applicable.

Wheelchair Skills Training

General training tips

- *Adjustment tips*
  - As was noted earlier with respect to adjustments that make it easier for the wheelchair casters to be transiently popped from the surface, the wheelchair type and setup influence the ease with which the wheelchair can be tipped backward into the full wheelie position. It is easier to achieve the wheelie position in a wheelchair that is less stable to begin with; this can be achieved by moving the rear axle position forward.
  - If RADs do not allow the wheelchair to be tipped back far enough, they need to be adjusted out of the way or removed. Even for RADs that do permit a wheelie to be performed, they may not be sufficiently stable to prevent a full rear tip. To check this, the tester or trainer can tip the occupied wheelchair until it is resting on the RADs. With a spotter behind the wheelchair, the wheelchair user should try to tip the wheelchair over backward, by reaching and leaning backward; if the wheelchair does not tip over, then the RADs can be considered effective.

Special considerations for manual wheelchairs operated by users (Version 1)

- The description that follows is for people using two hands for propulsion, but people who only have the use of one arm can perform wheelies in a similar way.
- The learner should be cautioned that most people require a total of 45–60 minutes of practice, spread over two to three sessions, to acquire and retain this skill.
- The sequence of phases trained is not critical. It is acceptable to start with the balance phase before proceeding to the takeoff phase, but the more natural sequence is described below.

- *Takeoff phase*
  - The learner will already have learned how to transiently pop the casters from the surface in earlier skills. It may be useful to review caster popping before proceeding to the full wheelie takeoff.
  - It may be useful to use simulation, having the trainer tip the wheelchair back into the balance position, to give the wheelchair user a sense of how much tilt will be needed.
  - If properly timed and the wheelchair is appropriately set up, the wheelchair user should require little force to achieve takeoff.
  - For the wheelie takeoff, many wheelchair users roll backward slowly, then quickly forward. This method is very effective and is to be preferred when the wheelchair user wishes to perform a wheelie

without moving forward at all. If using this method, the wheelchair user should start with the hands near the top center of the wheel (i.e., ~1:00 o'clock, using the clock analogy). The wheelchair user should try not to pause between rolling back and pushing quickly forward, otherwise he/she may not tip backward as easily.

o The method of only rolling the wheels forward is preferred because it can be used while the wheelchair is moving forward (as is occasionally necessary). The hands will need to start farther back on the wheels (i.e., ~11 o'clock) and slightly more force will be needed by the wheelchair user than for the backward–forward method.

o The forward motion that is common to both methods can be thought of as an action to get the base of support (the rear wheels) under the center of gravity (located near the lap).

o Some wheelchair users may find it easier if they lean back into the backrest to cause or help with the initial rear tip. However, skilled wheelie performers can achieve the wheelie position while maintaining an upright body position. Leaning or hunching forward is a natural tendency but will make it more difficult to achieve takeoff.

o Whichever method is used, the wheelchair user should progressively pop the casters higher and higher until he/she can tip backward far enough to reach and slightly overshoot the wheelie balance point. Once past the balance point, the wheelchair user should then pull back on the hand-rims to prevent tipping too far and to return to the balance point.

o If the wheelchair user is overshooting the balance point too vigorously, a learning exercise is for him/her to practice popping the casters up onto a small object (~5 cm high).

o If the wheelchair user is having difficulty getting tipped far enough backward to reach the balance point, he/she should push forward more forcefully. An alternative is to start the takeoff with the casters uphill or on a small level change although there needs to be room for the rear wheels to roll forward. If the problem is fear of tipping over backward, the wheelchair user can pop back onto the spotter then progress to a "self-save" (flexing the neck and trunk while pulling back vigorously on the hand-rims to bring the casters back to the floor). Once the learner is able to tip backward far enough to be caught by the spotter, in subsequent attempts he/she should gradually reduce the amount of overshoot until it is possible to self-save without the spotter's assistance.

o Although takeoff can usually be achieved with a single push, if the wheelchair has not been tilted back far enough with the first push, a second push before the casters return to the floor may be successful.

o Once the learner can consistently perform the wheelie takeoff, attention should be shifted to the balance phase.

- *Balance phase*
  - The wheelchair user does not need to use much force to maintain balance. It is preferable for the wheelchair user to keep a light grip on the wheels. It should be possible for the wheelchair user to slide the hands forward and backward on the hand-rims.
  - The wheelchair user should try to relax and remember to breathe.
  - During the early learning stage, some wheelchair users find it useful to isolate the variations of pitch from those of rear-wheel displacement (i.e., using the motor-learning principle of reducing the degrees of freedom). This can be done by reducing the extent to which the rear wheels can move (e.g., obstacles such as bricks or pieces of wood in front of and behind the rear wheels) (Figure 7.98). If the wheelchair is well set up and the wheelchair user has adequate strength, he/she may be able to push forward hard enough to tilt the wheelchair into the balance position. Otherwise, the trainer can tip the wheelchair back to the balance point while the wheelchair user rests his/her hands in the lap (Figure 7.99). For a wheelchair that is difficult for the trainer to tip backward (e.g., due to a low backrest, absence of push-handles, absence of tipping levers, excessive stability), the trainer can pull backward on the upper anterior chest with one hand or forearm or alternatively lift a forward section of the wheelchair frame. The trainer then turns over control to the wheelchair user by having the wheelchair user grasp the hand-rims (Figure 7.100). The trainer should then take his/her hands off the wheelchair and wheelchair user—it can be confusing to have two

Figure 7.98 Using bricks to immobilize the rear wheels.

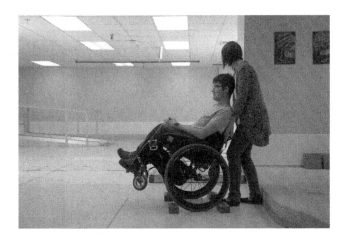

Figure 7.99 Trainer tilts the wheelchair into the wheelie position with the wheelchair user's hands in the lap.

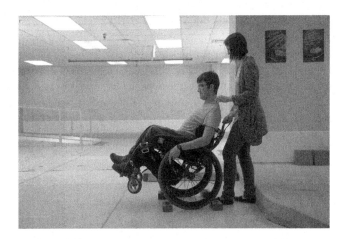

Figure 7.100 Wheelchair user places the hands on the hand-rims.

people attempt to maintain balance at the same time—and let the learner know ("*It's all you now*").

o Once the wheelchair user is in control with the rear wheels blocked, learning exercises include (i) having the wheelchair user experiment with the extent of tip (more and less than the ideal balance point, where the force to maintain position is minimal), (ii) leaning forward (which increases the amount of tip needed to be at the ideal balance point), (iii) using only two fingers and a thumb of each hand on the hand-rims, (iv) sliding the hands backward and forward on

269

Figure 7.101 Wheelchair user demonstrates a light touch by waving with one hand.

the hand-rims to find the ideal position, (v) holding on with only one hand while waving the other (Figure 7.101), and/or (vi) closing the eyes.

o Once these variations are mastered at the high–rolling resistance level, the barriers in front of or behind the rear wheels can be moved a few centimeters away by the wheelchair user or trainer (Figure 7.102). This allows a small amount of forward and backward movement of the rear wheels. At either extreme of movement, the wheelchair user can lean against the barriers. Once the wheelchair user is

Figure 7.102 With the bricks slightly apart, the rear wheels are free to move within limits.

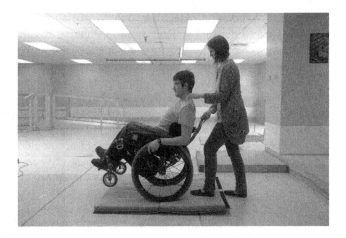

Figure 7.103 Practicing the stationary wheelie on a gym mat.

familiar with this, the barriers can be moved progressively farther away and removed.

o Once the wheelchair user has become comfortable with not spending too much time leaning on the barriers, the wheelchair can be moved to a surface with medium rolling resistance (e.g., on 5 cm of foam) (Figure 7.103). Here the takeoff and balance phases can be combined. The soft surface allows the learner to perform a "slow-motion" wheelie.

o Once this is mastered, the wheelchair can be moved to a low-rolling resistance surface (e.g., a tile floor).

o Once a basic wheelie can be performed on a low-rolling resistance surface, the learner can refine his/her skill by becoming more familiar with and practicing the two balance strategies that have been reported in the scientific literature.

  – *Proactive balance strategy*: In this strategy, analogous to balancing a meter stick on a finger, the wheelchair user keeps the wheels moving forward and backward over a small area. The wheelchair user should try to move the hands only between the 12:00 and 1:00 o'clock positions. This will allow a safety margin, so that the wheelchair user can react to a loss of balance in either direction. If the wheelchair user wants the wheels to move farther than the hand position permits, the hand-rims can be allowed to slide through the grip. It may be helpful to time the movement of the rear wheels to the breathing pattern while using the proactive balance strategy.

- *Reactive balance strategy*: The reactive balance strategy is analogous to the step strategy used in standing balance—if a standing person is pushed forward or backward hard enough that he/she would otherwise fall, the person steps forward or backward to bring the base of support under the displaced center of gravity. If the wheelchair user begins to tip too far forward, he/she should roll the rear wheels forward to return to the balance point ("*When you fall forward, push forward*"). If the wheelchair user imbalances backward, he/she should roll the rear wheels backward to reestablish balance ("*When you fall back, pull back*"). Even if past the point of no return and a full rear tip is imminent, the preferred strategy to minimize injury due to striking the back of the head on the ground is for the wheelchair user to pull backward forcefully on the hand-rims and flex the neck until the back hits the ground. Although some authorities advise wheelchair users to use one or both hands on the knees during a rear fall, to prevent the knees from striking the wheelchair user in the face, as the saying goes "a broken nose is preferable to a broken skull." Rear falls will be practiced later as part of the "gets from ground into wheelchair" skill. The reactive balance strategy will be used later, to deliberately move the wheelchair forward when beginning the "descends high curb in wheelie position" and "descends steep incline in wheelie position" skills.

• *Landing phase*
  o To land, the wheelchair user pulls back on the wheels, or leans forward to gently bring the front wheels to the ground.

• *Progression*
  o Once the full wheelie can be performed with the spotter nearby, the wheelchair user can practice performing the stationary wheelie with variations (e.g., with the spotter progressively farther away, with low lighting, while multi-tasking).

• *Variations*
  o During the balance phase, the wheelchair user can lean forward or place a knapsack on the lap or footrests to increase the caster height needed for the wheelie position. The wheelchair user can practice this by placing the casters on different height targets (e.g., pylons, steps).

Special considerations for manual wheelchairs operated by caregivers (Version 2)

- As noted earlier in the "rolls on soft surface" skill, to achieve a care-giver-induced wheelie, the caregiver should pull back on the push-handles, with one foot pushing down on a tipping lever, to tip the wheelchair back to the balance point.
- Once in the wheelie balance position, only minimal force is needed by the caregiver to maintain balance.
- To lower the wheelchair to the horizontal position, the caregiver should put one foot on the tipping lever at the back of the wheelchair to keep the wheelchair from pitching forward too abruptly.

Special considerations for powered wheelchairs operated by users (Version 3)

- Not applicable.

Special considerations for powered wheelchairs operated by caregivers (Version 4)

- Not applicable.

Special considerations for scooters operated by users (Version 5)

- Not applicable.

7.34 Turns in place in wheelie position

Versions applicable

Version	Wheelchair Type	Subject Type	Applicable
1	Manual	Wheelchair user	✓
2		Caregiver	✓
3	Powered	Wheelchair user	
4		Caregiver	
5	Scooter	Wheelchair user	

Skill level
- Advanced.

Description
- In the wheelie position, the subject turns the chair in place, both to the left and right.

Rationale
- Wheelchair users often encounter situations in which they need to perform a wheelie to make a tight turn. The area needed on the support surface (the "footprint") is less in the wheelie position than when all wheels are on the surface.

Prerequisites
- "Performs stationary wheelie" skill.

Spotter considerations
- Spotter starting position: Behind the wheelchair, holding onto the spotter strap.
- Risks requiring spotter intervention: Rear tip if the wheelchair user overshoots on takeoff or loses balance.

Wheelchair Skills Test (WST)

Equipment
- As for the "performs stationary wheelie" skill.

Starting positions
- As for the "performs stationary wheelie" skill.

Instructions to subject
- *"Get the wheelchair into the wheelie position."*
- *"Now, keeping the chair within this square* (indicate it), *turn the wheelchair around until it is facing the opposite direction."*
- *"Now turn the chair in the other direction* (indicate it) *until it is back where you started."*

Capacity criteria
- As for the general scoring criteria, with the clarifications below.
- As for the "turns in place" and "performs stationary wheelie" skills, except as below.
- The subject is permitted to return the casters to the floor between the turns to the left and right.
- A "fail" score should be awarded if the subject fails the "performs stationary wheelie" skill.

Special considerations for manual wheelchairs operated by users (Version 1)
- None.

Special considerations for manual wheelchairs operated by caregivers (Version 2)
- The caregiver must keep his/her feet within the boundaries, as for the "turns in place" skill.

Special considerations for powered wheelchairs operated by users (Version 3)
- Not applicable.

Special considerations for powered wheelchairs operated by caregivers (Version 4)
- Not applicable.

Special considerations for scooters operated by users (Version 5)
- Not applicable.

Wheelchair Skills Training

General training tips
- This skill is a combination of the "turns in place" and "performs stationary wheelie" skills.
- Although the footprint for a wheelie turn in place is small, the space needed above the ground may be at least as great as with all wheels on the ground. The subject should be careful not to let the elevated feet hit any external object.

Special considerations for manual wheelchairs operated by users (Version 1)
- As for the "turns in place" and "performs stationary wheelie" skills.

- *Progression*
  - The wheelchair user should begin with small angular displacements around the yaw axis (the vertical axis between the two rear wheels) that do not require that the hands be repositioned.
  - The wheelchair user should then progress to larger displacements that require the hands to be repositioned, using several steps to get all the way around to 180°.

- o Some wheelchair users may be able to get all of the way around to 180° (or beyond) in a single movement (the so-called "snap turn") by allowing the hand-rims to slide through the fingers.
- o The wheelchair user can practice on progressive smaller areas of support.
- o The wheelchair user can practice on a soft surface.

Special considerations for manual wheelchairs operated by caregivers (Version 2)
- As for the "turns in place" and "performs stationary wheelie" skills.

Special considerations for powered wheelchairs operated by users (Version 3)
- Not applicable.

Special considerations for powered wheelchairs operated by caregivers (Version 4)
- Not applicable.

Special considerations for scooters operated by users (Version 5)
- Not applicable.

7.35 Descends high curb in wheelie position

Versions applicable

Version	Wheelchair Type	Subject Type	Applicable
1	Manual	Wheelchair user	✓
2		Caregiver	✓
3	Powered	Wheelchair user	
4		Caregiver	
5	Scooter	Wheelchair user	

Skill level
- Advanced.

Description
- In the wheelie position, the subject descends a high curb in the forward direction.

Rationale
- Large level changes (e.g., curbs, steps) are common obstacles for wheelchair users. Using a wheelie to descend a level change in the forward direction allows the wheelchair user to maintain forward movement and to see any dangers that lie ahead. Also, the wheelie position prevents the footrests from making contact with the lower level, which can decelerate the wheelchair and cause a forward tip or fall from the wheelchair.

Prerequisites
- "Performs stationary wheelie" skill.

Spotter considerations
- Spotter starting position:
  - For a single spotter behind the wheelchair, if no seat belt is used one of the spotter's hands should be placed near a push-handle and the other hand in front of the wheelchair user's shoulder.
  - For a single spotter behind the wheelchair, if a seat belt is used, both of the spotter's hands may be placed near the push-handles of the wheelchair.
  - If using two spotters, the second spotter should stand beside and below the curb.
- Risks requiring spotter intervention:
  - Rear tip.
  - Forward tip or fall.
  - Sideways tip if one wheel drops off the upper level before the other.

Wheelchair Skills Test (WST)

Equipment
- As for the "descends high curb" skill.

Starting positions
- As for the "descends high curb" skill.

Instructions to subject
- Screening questions (*"Can you do it? How?"*) are strongly recommended before the subject is allowed to proceed to the objective testing of this skill.
- *"Get your wheelchair into the wheelie position."*
- *"Now, staying in the wheelie position, move forward down the curb under control."*

Capacity criteria
- As for the general scoring criteria, with the clarifications below.
- Generally, as for the "descends high curb" and "performs stationary wheelie" skills.
- A "pass" should be awarded if:
  - o The subject achieves a controlled wheelie on the upper level, approaches the curb by moving forward in this position and then lowers the rear wheels to the lower level under control with the rear wheels striking the floor before the casters.
  - o A transient caster pop may be used instead of a full wheelie, as long as the casters do not strike the floor before the rear wheels.
- A "fail" score should be awarded if the subject fails the "performs stationary wheelie" skills.

Special considerations for manual wheelchairs operated by users (Version 1)
- None.

Special considerations for manual wheelchairs operated by caregivers (Version 2)
- None.

Special considerations for powered wheelchairs operated by users (Version 3)
- Not applicable.

Special considerations for powered wheelchairs operated by caregivers (Version 4)
- Not applicable.

Special considerations for scooters operated by users (Version 5)
- Not applicable.

Wheelchair Skills Training

General training tips

- This skill is a combination of the "descends high curb" and "performs stationary wheelie" skills.

Special considerations for manual wheelchairs operated by users (Version 1)

- As for the "descends high curb" and "performs stationary wheelie" skills.
- The forward full wheelie method is the preferred method for the descent of a large level change, but it requires good wheelie skills (Figure 7.104a through d). The wheelchair user should get into the wheelie position

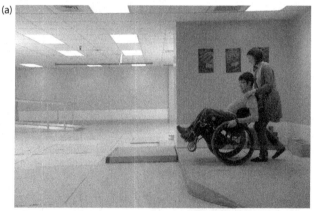

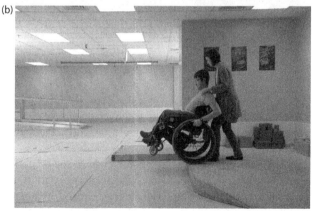

Figure 7.104 For the "descends high curb in wheelie position" skill: (a) The wheelchair user gets into the wheelie position away from the curb edge. (b) The wheelchair is moved forward to the curb edge. *(Continued)*

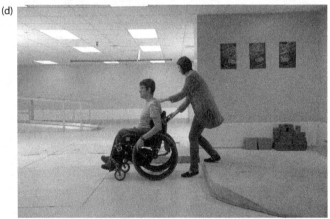

Figure 7.104 (Continued) (c) The rear wheels are lowered to the lower level, and (d) the casters land after the rear wheels.

away from the edge of the curb. The wheelchair user should roll forward to the edge of the curb in the wheelie position, staying as square as possible to the edge.

- To move forward on the level above the curb in the wheelie position, the wheelchair user should allow the wheelchair to begin to fall ("dip") slightly forward, and then should roll the rear wheels forward to catch up. This is like the reactive balance strategy described in the performs "stationary wheelie" skill, but the initial imbalance is intentional. It may be easier to begin with very small steps forward.
- When moving forward or backward in the wheelie position, to initiate the dip, the wheelchair user can move the head or lean slightly in

the direction toward which he/she wishes to move. Alternatively, the wheelchair user can initiate the dip by pushing the wheels slightly in the opposite direction. The wheelchair user should be encouraged to take his/her time to achieve control and to move slowly. The wheelchair user should grip the wheels lightly, giving a light push on the wheels to move them forward, letting the hand-rims slide through the fingers. In catching the rear wheels up to the center of gravity after the dip, there is no need for the wheelchair user to catch up completely. By undershooting slightly, the wheelchair user can initiate the next dip.

- After initiating the forward dip to move the rear wheels over the edge of the curb, the wheelchair user should quickly slide the hands backward from the 1:00 o'clock to the 11 o'clock position (clock analogy), so that he/she can firmly grip the hand-rims long enough for the rear wheels to drop all the way to the lower level. As slowly as possible, the wheelchair user should lower the rear wheels from the upper to the lower level, pulling backward to slow the descent. The wheelchair user should let the rear wheels hit the lower level before the casters. As soon as the rear wheels touch the ground, the momentum should bring the casters down, but the wheelchair user should lean forward as well.

- *Progression*
  - Moving forward in the wheelie position should be practiced first on a level surface away from the curb, proceeding up to a line that represents the top edge of the curb. A strip of bubble wrap can be used to provide audible feedback as to when the line is reached.
  - As a learning exercise, the forward and backward "dip-and-roll" processes can be practiced against resistance (e.g., on a soft surface, up an incline, over a threshold, up a 5 cm curb). The dip needs to be accentuated in such circumstances.
  - The skill should be practiced first on a low curb, increasing the height of the curb as skill and confidence increase.

- *Variation*
  - The wheelchair user can land on the lower level and maintain the wheelie position rather than allowing the casters to land, either maintaining balance or leaning back against the curb rise. This is useful where there is little space for the casters to land, such as on a series of widely spaced stairs.

Special considerations for manual wheelchairs operated by caregivers (Version 2)
- As for the "descends high curb" and "performs stationary wheelie" skills.

Special considerations for powered wheelchairs operated by users (Version 3)
- Not applicable.

Special considerations for powered wheelchairs operated by caregivers (Version 4)
- Not applicable.

Special considerations for scooters operated by users (Version 5)
- Not applicable.

7.36 Descends steep incline in wheelie position

Versions applicable

Version	Wheelchair Type	Subject Type	Applicable
1	Manual	Wheelchair user	✓
2		Caregiver	✓
3	Powered	Wheelchair user	
4		Caregiver	
5	Scooter	Wheelchair user	

Skill level

- Advanced.

Description

- In the wheelie position, the subject descends a steep incline.

Rationale

- Descending a steep incline in the forward direction in the wheelie position lessens the problem of loss of traction (affecting braking and control) when the uphill wheels become unloaded. This technique also reduces the likelihood of forward tips or digging the footrests into the floor at the transition between the bottom of the incline and the level surface. For very steep inclines, this technique may be the only way to get down the incline without tipping over.

Prerequisites

- "Performs stationary wheelie" skill.

Spotter considerations

- As for "descends steep incline."

Wheelchair Skills Test (WST)

Equipment

- As for "descends steep incline."

Starting positions

- As for "descends steep incline."

Instructions to subject

- *"Get your wheelchair into the wheelie position."*
- *"Now, staying in the wheelie position, move down the ramp under control, and stop when you reach the floor at the bottom."*

Capacity criteria

- As for the general scoring criteria, with the clarifications below.

- Generally, as for the "descends steep incline" and "performs stationary wheelie" skills.
- A "pass" should be awarded if:
  - The subject achieves the wheelie position on the platform above the incline, proceeds down the incline with the chair under control and brings the wheelchair to a stop in the space available at the bottom of the ramp.
  - The casters may be brought to the surface as soon as the rear wheels have reached the level surface.
- A "fail" score should be awarded if the subject fails the "performs stationary wheelie" skills.

Special considerations for manual wheelchairs operated by users (Version 1)
- None.

Special considerations for manual wheelchairs operated by caregivers (Version 2)
- None.

Special considerations for powered wheelchairs operated by users (Version 3)
- Not applicable.

Special considerations for powered wheelchairs operated by caregivers (Version 4)
- Not applicable.

Special considerations for scooters operated by users (Version 5)
- Not applicable.

Wheelchair Skills Training
General training tips
- This skill is a combination of the "descends steep incline" and "performs stationary wheelie" skills.

Special considerations for manual wheelchairs operated by users (Version 1)
- The wheelchair user usually achieves the wheelie position on the level at the top of the incline (Figure 7.105).
- Then, he/she moves forward onto the incline.
- When initially moving onto the incline, the wheelchair user may be startled to feel as though the wheelchair is tilting farther backward.
- Once on the incline, facing downhill (Figure 7.106), the wheelchair user should let the hand-rims run smoothly through the hands to control the wheelchair's speed, direction, and pitch angle. Letting the hand-rims run more quickly through the hands will allow the wheelchair to pitch (tilt)

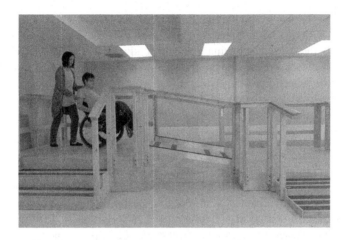

Figure 7.105 For the "descends steep incline in wheelie position" skill, the wheelchair user gets into the wheelie position away from the incline transition.

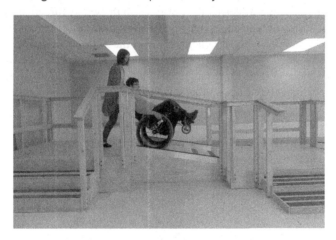

Figure 7.106 The wheelchair user uses the hand-rims of the wheelchair to control pitch, speed, and direction.

farther back. Slowing the rate at which the hand-rims slide through the fingers will cause the wheelchair to pitch forward.
- The subject should have the casters touch down shortly after the rear wheels reach the level surface at the bottom of the incline.

- *Progression*
    - o Moving forward and backward in the wheelie position will already have been practiced for the "descends high curb in the wheelie position" skill.

- o Some people find it easier to perform this skill on steep inclines than slight inclines.
- o If the learner is having difficulties advancing the wheelchair from the level section onto the incline, he/she may find it easier to get into the wheelie position while already on the incline, facing sideways (see variation below) and then turn the wheelchair downhill.

- *Variations*
  - o As for "descends steep incline" skill (e.g., steering a slalom path [Figure 7.107], starting, stopping).
  - o When stopped facing downhill in the wheelie position, the sensation is similar to that felt while leaning back on a barrier, as when learning the balance phase of the "performs stationary wheelie" or the "tilt-rest" skill.
  - o The subject can achieve wheelie takeoff while on the incline. This is useful when an unexpected obstacle is encountered. If the wheelchair user is facing downhill, more force is needed for takeoff (because the wheelchair is pre-tilted in the wrong direction) and the wheelchair may accelerate rapidly downhill.
  - o On steep or slippery inclines, or if the wheelchair has too much rear stability, there may not be enough rear-wheel traction to allow wheelie takeoff while facing downhill. In such situations, the wheelchair can be turned so that it is facing across the hill or even uphill. This will place more weight on the rear wheels and avoid runaway. Once in the wheelie position, a wheelie turn-in-place will allow the wheelchair user to proceed down the incline.

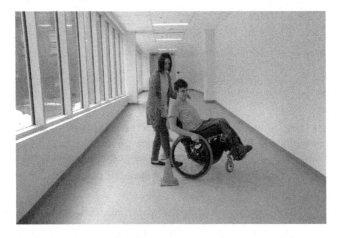

Figure 7.107 Descending an incline through pylons in the wheelie position.

Special considerations for manual wheelchairs operated by caregivers (Version 2)
- As for the "descends steep incline" and "performs stationary wheelie" skills.
- Descending an incline forward in the wheelie position is comfortable for the wheelchair user with no risk of falling out of the wheelchair. Also, the caregiver has the advantage of being able to see where he/she is going.

Special considerations for powered wheelchairs operated by users (Version 3)
- Not applicable.

Special considerations for powered wheelchairs operated by caregivers (Version 4)
- Not applicable.

Special considerations for scooters operated by users (Version 5)
- Not applicable.

7.37 Gets from ground into wheelchair

Versions applicable

Version	Wheelchair Type	Subject Type	Applicable
1	Manual	Wheelchair user	✓
2		Caregiver	✓
3	Powered	Wheelchair user	✓
4		Caregiver	✓
5	Scooter	Wheelchair user	✓

Skill level
- Advanced.

Description
- The wheelchair user gets from the ground into the wheelchair.

Rationale
- This skill is useful when recovering from a fall or from an occasion when the wheelchair user is on the ground for another reason.

Prerequisites
- None.

Spotter considerations
- Spotter starting position:
  - If there is a single spotter, he/she should be near the wheelchair, in a position to prevent the wheelchair from tipping over or to prevent the subject from falling to the ground.
  - If two spotters are used, one spotter should focus on the wheelchair user and the other spotter on preventing the wheelchair from sliding or rolling away. However, the second spotter should not touch the wheelchair unless it is necessary to intervene.
- Risks requiring spotter intervention: Rear, forward, or sideways tip or fall.

Wheelchair Skills Test (WST)
Equipment
- Smooth level surface: It is permissible to use a thin mat to protect the skin and to avoid dirt.
- No external aids (e.g., the transfer bench, stairs) may be used.

Starting positions
- Wheelchair user: Seated or lying on the ground, out of the wheelchair. If the transfer to the ground cannot be achieved independently, the tester can assist the wheelchair user out of the wheelchair.
- Wheelchair: Within reach, with the wheel locks unlocked.

Instructions to subject
- Screening questions (*"Can you do it? How?"*) are strongly recommended before the subject is allowed to proceed to the objective testing of this skill.
- *"Get into the wheelchair."*

Capacity criteria
- As for the general scoring criteria, with the clarifications below.
- A "pass" should be awarded if:
  - The subject gets onto the wheelchair seat, ready to roll away.
  - The subject removes the seat cushion as part of his/her technique, it is required that the cushion be picked up but it is not necessary to get the cushion back under the buttocks. This is in recognition that the wheelchair user is usually able to go to another sitting surface and transfer out of the wheelchair to replace the cushion. The "level transfer" was assessed earlier.
- A "fail" score should be awarded if the subject does not describe a safe and effective method.
- A "testing error" score may be awarded if the transfer to the ground cannot be achieved independently or with the assistance of the tester.

Special considerations for manual wheelchairs operated by users (Version 1)
- None.

Special considerations for manual wheelchairs operated by caregivers (Version 2)
- The caregiver may receive assistance from the wheelchair user in performing the skill. This is an exception to the general rule that the wheelchair user should not assist when the caregiver is being assessed alone because it is not a reasonable expectation that a single caregiver could carry out this skill alone.

Special considerations for powered wheelchairs operated by users (Version 3)
- None.

Special considerations for powered wheelchairs operated by caregivers (Version 4)
- None.

Special considerations for scooters operated by users (Version 5)
- None.

Wheelchair Skills Training

General training tips
- To get from the wheelchair to the ground, the trainer may assist, using a lift if available. If it is a feasible goal of the wheelchair user to be able

to independently get to the ground, then the steps described below for getting from the ground to the wheelchair can be reversed.

- After a fall, unless there is some immediate danger, the wheelchair user and/or caregiver should take time to assess whether there has been any injury or damage to the wheelchair or occupant before getting back into the wheelchair.
- There are a number of techniques that wheelchair users can use to get safely back into their wheelchairs from the ground, the variations reflecting differences in the nature of the wheelchair users' impairments and wheelchair characteristics. Only a few of the more commonly used techniques will be described. There is no available literature as yet supporting the superiority of one technique over the others. The trainer and wheelchair user may wish to try the variations before selecting the one that will be used in most circumstances.

Special considerations for manual wheelchairs operated by users (Version 1)

- *Fall practice*
  - Getting from the wheelchair onto the ground is an opportunity to practice and/or discuss safe falls.
  - Generally, regardless of the fall direction, the wheelchair user should not reach out toward the ground with an arm because even an otherwise minor arm injury can have major functional consequences for a person who uses that arm for mobility and transfers. However, some wheelchair users with low backrests, long arms, and good flexibility can prevent full rear or sideways tips with a gentle push on the ground.
  - Rear falls can be safely practiced. The trainer should first lower the wheelchair user onto an elevated gym mat, with the wheelchair user's neck flexed and hands pulling on the hand-rims. Failure to hold onto the hand-rims will result in the rear wheels of the wheelchair rolling rapidly forward ("submarining"). The wheelchair user can then progress to real falls onto an elevated mat, the height of which can be progressively lowered. As described earlier, if a rear fall seems imminent, the wheelchair user should flex the neck and pull backward as forcefully as possible on the hand-rims. In addition to preventing submarining, the rate of rear tip will be decreased and the arms will act as shock absorbers when the wheelchair strikes the floor. Immediately after hitting the ground, the wheelchair user can use the hands or forearms to prevent the knees from striking the face.
  - There is no safe and practical way to practice forward or sideways falls. However, they should at least be discussed. During a forward fall, the wheelchair user should twist to one side and try to roll

sideways after striking the ground, protecting the head with the hands. During a sideways fall, the wheelchair user should lean away from the direction of tip, pulling vigorously on the uphill armrest or hand-rim.

- *Out-of-wheelchair approach for getting from the ground to the wheelchair*
  - First, the wheelchair should be righted, the casters should be oriented so that they are trailing forward, the wheel locks should be applied and, unless they will be used as an intermediate sitting surface, the footrests should be moved out of the way if possible.
  - The wheelchair user should be in the sitting position in front of the wheelchair, with the hips and knees flexed as much as possible.
  - The wheelchair user can use the seat cushion to increase the height of the floor and to lower the height of the wheelchair seat. After getting up onto the seat, the cushion can be placed back under the buttocks by rolling to a transfer surface that is the same height as the wheelchair seat and transferring out of the wheelchair.
  - The wheelchair user sitting sideways at the front of the wheelchair can lift the buttocks with one arm on the seat and one on the ground. This approach is similar to a sideways level transfer (discussed earlier). Moving the head in the direction opposite to the direction to the hips is useful (i.e., move the head down when moving the hips up). This technique can also be performed with the wheel locks off. As the wheelchair user lifts the buttocks off the floor, he/she can simultaneously pull the wheelchair under the buttocks.

  - *Variations*
    - The wheelchair user with his/her back facing the front of the wheelchair can lift the buttocks with both hands on the seat or front rigging at the same time. The footrests can be used as an intermediate level between the ground and the wheelchair seat, if they are wide enough and if sitting on them does not tip the wheelchair forward.
    - The wheelchair user can move progressively from the floor to a foot stool, a bench, and finally to the wheelchair seat. The number of steps can be gradually reduced.
    - Some wheelchair users may find it easier to face the wheelchair, getting up onto the knees before moving up to the seat level and twisting into the forward-facing position.
    - If the wheelchair user has the use of his/her legs, he/she can use the wheelchair to help get up onto his/her feet, then pivot and sit down.

- If there is another stable object nearby (e.g., a chair, low table), the wheelchair user can put one hand on the object and the other hand on the wheelchair seat.

- *Stay-in-wheelchair approach*
  - o Some wheelchair users are able to right themselves while remaining in the wheelchair.
  - o To train someone to perform this technique, the wheelchair user can start on a surface partway between seat height and the ground, with the wheelchair on its back (as would be the case after practicing a fall backward onto an elevated mat, as described above).
  - o The wheelchair user should pull on the rear wheels to get the buttocks firmly against the wheelchair seat.
  - o The wheelchair user may let the knees bend over the front of the seat.
  - o The wheel lock should be applied on the side of the stronger arm.
  - o The wheelchair user turns the trunk to the other side and uses the forward (stronger) hand to grab the hand-rim of the rear wheel on the unlocked side, with the hand as far forward as possible.
  - o The wheelchair user then reaches with the other hand to the surface on which the backrest of the wheelchair rests.
  - o The wheelchair user simultaneously and vigorously pushes with the floor hand and pulls with the hand-rim hand. This step is repeated as necessary, moving the floor hand progressively forward on the surface and the hand-rim backward until upright.

Special considerations for manual wheelchairs operated by caregivers (Version 2)
- The caregiver can assist the wheelchair user by helping to position and stabilize the wheelchair.
- The caregiver should try to avoid bending and twisting his/her back at the same time and should lift with bent knees.
- If tipping the wheelchair upright from the fully rear-tipped position, locking the wheel locks will keep the wheelchair from rolling forward (submarining).
- A single caregiver may have difficulty in performing this skill without the help of the wheelchair user and/or a second caregiver. A mechanical lift or a team of people are recommended when lifting from the floor.
- If the caregiver is large and strong and the wheelchair user is light, the caregiver may be able to safely lift the wheelchair user from the side, with one arm around the back and under the arms and the other arm under the bent knees.

- If there are two caregivers, they may pick up the wheelchair user together. This can be done in two ways:
  - One option is to have one caregiver behind the wheelchair user, holding the wheelchair user's arms by reaching under the upper arms and grasping the folded forearms. The other caregiver lifts with his/her hands behind the wheelchair user's knees.
  - The other option is for the two caregivers to be on opposite sides of the wheelchair user, each with one arm under one of the wheelchair user's arms and around the back and the other arm under the wheelchair user's bent knees.
- If a third caregiver is available, he/she can help with the legs or manage the wheelchair. In some circumstances, it may be practical to move the wheelchair under the lifted wheelchair user rather than moving the wheelchair user to the wheelchair.

Special considerations for powered wheelchairs operated by users (Version 3)
- If falling backward in a powered wheelchair, the wheelchair user should tuck the chin and pull himself/herself vigorously forward using the armrests or seat. This should be discussed but cannot be easily practiced.
- After a fall, the power should be turned off. The power should be off while the getting-up skill is being practiced.
- Those involved should check to be sure that there is no spilled battery acid.

Special considerations for powered wheelchairs operated by caregivers (Version 4)
- As for Versions 2 and 3.

Special considerations for scooters operated by users (Version 5)
- As for Version 3.

7.38 Ascends stairs

Versions applicable

Version	Wheelchair Type	Subject Type	Applicable
1	Manual	Wheelchair user	✓
2		Caregiver	✓
3	Powered	Wheelchair user	
4		Caregiver	
5	Scooter	Wheelchair user	

Skill level
- Advanced.

Description
- The wheelchair user and the wheelchair get from the bottom of a set of stairs to the top.

Rationale
- Although alternative means of getting from a lower to a higher level are often present (e.g., using a ramp, elevator), stairs may sometimes be the only option. Although exceptional manual wheelchair users can accomplish this skill alone while sitting in the wheelchair, it is not recommended due to the ergonomic arm stresses involved. Getting out of the wheelchair or using caregivers to assist with stair ascent is a more reasonable approach. This skill is not applicable for most powered wheelchairs and scooters.

Prerequisites
- None.

Spotter considerations
- Spotter:
  - If a single spotter is used, he/she should be below the wheelchair with one hand near or holding a fixed part of the wheelchair and the other hand on a stair handrail. If holding a wheelchair part, it is important to avoid assisting or interfering with the performance of the task unless deliberately intervening.
  - If two spotters are available, one should be above and one below the wheelchair.
- Risks requiring spotter intervention:
  - Forward or rear tip or fall.
  - Runaway down the stairs.

Wheelchair Skills Test (WST)

Equipment
- There should be at least three stairs, with the following approximate dimensions—18 cm rise, 28 cm run, and width of at least 1.2 m. (In describing a set of stairs, one refers to the horizontal and vertical dimensions as the "run" and "rise," respectively.) Although three stairs are not many, they are representative of the skills needed for a full flight of steps.
- Rails should be available on both sides, at a height above the steps of about 90 cm. The rails should extend horizontally beyond the upper and lower stair boundaries by 30 cm or more.
- The set of stairs should end at the upper end on a level surface or platform that is at least 2 m² square. A lip around the open edges of the platform is recommended.
- No external aids (e.g., stair lift) may be used.

Starting positions
- Wheelchair: Facing the stairs at least 0.5 m from the bottom stair.

Instructions to subject
- Screening questions (*"Can you do it? How?"*) are strongly recommended before allowing the subject to proceed with objective testing.
- *"Get yourself and the wheelchair up the stairs."*

Capacity criteria
- As for the general scoring criteria, with the clarifications below.
- A "pass" should be awarded if any effective and safe technique is permitted, as long as at least three stairs are completed.
- A "pass with difficulty" should be awarded if the wheelchair user overexerts him/herself.

Special considerations for manual wheelchairs operated by users (Version 1)
- None.

Special considerations for manual wheelchairs operated by caregivers (Version 2)
- As for the general scoring criteria, with the clarifications below.
- A "pass" should be awarded if:
  - The caregiver receives assistance from the wheelchair user in performing the skill.
- A "pass with difficulty" should be awarded if the caregiver overexerts him/herself.

Special considerations for powered wheelchairs operated by users (Version 3)
- Not applicable.

Special considerations for powered wheelchairs operated by caregivers (Version 4)
- Not applicable.

Special considerations for scooters operated by users (Version 5)
- Not applicable.

Wheelchair Skills Training

General training tips
- Alternative routes (e.g., ramps, elevators) to get to the upper level should be sought wherever possible.
- With the exception of the initial preparation for the first step and concluding the task after ascending the last step, the same technique is used for each step.
- Safety is of particular importance, given the consequences of a loss of control.

Special considerations for manual wheelchairs operated by users (Version 1)
- There are a variety of methods, the choice of which depends upon the characteristics of the wheelchair user (e.g., strength, flexibility, ability to use the legs) and the stairs.

- *Out of wheelchair, on buttocks*
  - A buttocks protector is recommended.
  - The wheelchair should be positioned next to the stairs, in a way similar to how the wheelchair would be positioned for the "level transfer" skill.
  - The wheelchair user transfers from the wheelchair to the second or third step, usually using a standing- or a crouch-pivot method. The stair handrail may be used.
  - The wheelchair may be brought up to the top the stairs by the wheelchair user or by an assistant. If bringing the wheelchair up the stairs himself/herself, the wheelchair user should pull the wheelchair up by facing it downhill, and tipping it back fully. The wheelchair user should push straight down with one hand on the wheelchair's push-handles that are resting on a step, to keep the wheelchair from rolling or sliding down the stairs.
  - As the wheelchair user moves up each step, he/she should flex the neck and hips and push down with the arms and feet to bring the buttocks up and back onto the next higher step. Then, the hands, feet, and wheelchair are moved up to the next step.
  - At the top of the stairs, this final phase is the same as for the "gets from ground into wheelchair" skill.

- *Variations*
  - *Out of wheelchair, on hands and knees*
    - As for the buttocks approach above, but facing up the stairs and using a crawling action.
  - *In wheelchair*
    - Although this technique is not recommended for wheelchair users acting alone, because of the long-term consequences of the stresses placed on the shoulders, the following tips are provided for the exceptional wheelchair user who wishes to acquire this skill for the unusual occasion when it would be helpful.
    - The RADs (if any) should be repositioned to allow the rear wheels to contact the first stair and to permit the wheelchair to tip backward sufficiently.
    - The starting position is with the wheelchair user in the wheelchair, with the seat belt (if any) on.
    - The wheelchair should be backed up to the lowest step, closest to the handrail on the side of the stronger arm.
    - The wheelchair user reaches back as far as he/she can with the stronger arm and grabs the handrail with the palm facing up.
    - By pulling on the handrail, the wheelchair user tilts the wheelchair back but not too far. The location of the combined center of gravity of the wheelchair user and the wheelchair is a key factor. If the center of gravity is behind the rear-wheel axles, the rear wheels will tend to submarine (i.e., move away from the vertical portion [the "rise"] of the step) if not prevented by the wheelchair user's or caregiver's hands on the hand-rims. If the center of gravity is in front of the rear-wheel axles, the rear wheels will tend to move backward, toward the vertical portion of the step (which is where they need to be to roll the rear wheels up the step).
    - The wheelchair user uses the hand on the stair handrail to pull while using the other hand on the hand-rim (starting well forward) to roll the rail-side wheel up the step.
    - Because both hands are acting on the same side of the wheelchair, the wheelchair will tend to turn toward the handrail. The wheelchair should be squared-up (i.e, both rear wheels against the step rise) before each new stair is attempted.
    - At the top of the stairs, the casters should not be brought down until there is surface to support them.
- *Escalators*
  - Escalators that are wide enough and are not excessively steep can be safely managed in a manual wheelchair. Permission should be obtained before practicing on escalators in public places.

- o To ascend an escalator, the wheelchair user approaches the lower end in the forward direction slowly, grasps both or one moving handrail and allows the wheelchair to be pulled onto the escalator.
- o The wheelchair will settle itself into a stable position. The wheelchair user should lean forward until on the level at the top.
- o The major difficulty comes at the top, where there is usually a lip that will stop the wheelchair or cause it to tip forward. To prevent this, the wheelchair user should lean well back without tipping the wheelchair over, still holding onto the handrails.
- o A spotter at the top can help to pop the casters over the lip until the wheelchair user has mastered this on his/her own.

- *Progression*
  - o It is useful to have stairs with a variety of runs and rises to permit gradual progression. The wheelchair user can use a curb first, if there is a rail beside it, as an example of a single step.
  - o It is reasonable to start with the caregiver-assisted versions of this skill. Caregivers can apply upward rolling forces to one or both rear wheels to assist in getting up the stair and to prevent the rear wheel on the side away from the handrail from moving away from the stair rise (Figure 7.108).
  - o If the staircase is curved, there is more "run" on the outside of the curve, so squaring-up requires more care.

Special considerations for manual wheelchairs operated by caregivers (Version 2)
- As for the section above, in many respects.

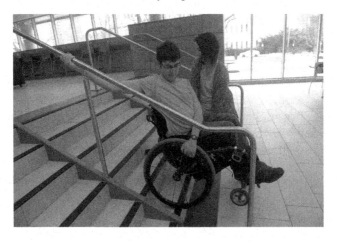

Figure 7.108 Ascending stairs seated in the wheelchair with the assistance of one caregiver.

- *Wheelchair user in the wheelchair*
  - If more than one caregiver is involved, as should usually be the case, the wheelchair user or one of the caregivers should by agreement take the lead in coordinating the timing (e.g., to the count of "ready, set, go" for each step).
  - The starting position is with the wheelchair user in the wheelchair, with the seat belt (if any) on. It can be helpful to remove the footrests.
  - The wheelchair should be backed up to the lowest step with the rear wheels firmly against the step rise.
  - The wheelchair user may place his/her hands on the rear wheels or the stair handrails, assisting to the extent possible but keeping his/her hands out of the way of the caregiver's hands.
  - Although not recommended because of the stresses involved, a single strong caregiver can help a light wheelchair user in a light wheelchair up the stairs from behind (uphill), tipping the wheelchair back beyond the balance point and rolling it up one step at a time.
  - Alternatively, if only a single caregiver is available and the wheelchair user is able to assist, then the caregiver can provide some of the needed force from downhill (e.g., rolling the non-rail-side wheel up the step while the wheelchair user pulls on the stair handrail with one hand and the rail-side hand-rim with the other hand) as described above in Version #1.
  - With two caregivers, one of the caregivers can be positioned uphill and pull on the push-handles while the other caregiver is below and pushes on the wheelchair frame (Figure 7.109).

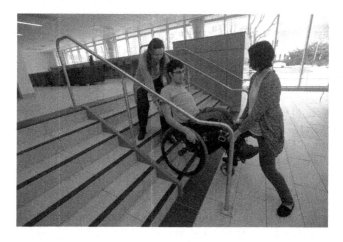

Figure 7.109 Ascending stairs seated in the wheelchair with the assistance of two caregivers.

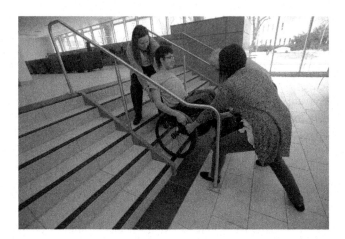

Figure 7.110 Ascending stairs seated in the wheelchair with the assistance of three caregivers.

- o If the wheelchair user cannot physically assist much, ideally there should be three caregivers available (Figure 7.110). One caregiver positions him/herself above, pulling on the push-handles, but not too forcefully because the awkward positioning could lead to injury of the caregiver's back. This uphill caregiver is turned slightly to one side, with one foot on the stair above the rear wheels and the other on the next higher stair. The primary role of the uphill caregiver is to control the degree of rear tilt, which should be ahead of the balance point as noted earlier. The uphill caregiver can tell where the center of gravity is relative to the balance point by whether the push-handles are pushing back or pulling forward (as they should be). If the wheelchair has a low backrest or no push-handles, one hand can be placed on the wheelchair user's upper anterior chest to control the extent of tilt. The two downhill caregivers are below the wheelchair. Each uses the inside hand (closest to the midline of the wheelchair) to hold the frame of the wheelchair, not a part (e.g., a footrest) that could come off. The outside hand is placed on the hand-rim of the rear wheel and is used to roll the wheel up onto the next step. The outside hand begins at about the horizontal position and moves up to the vertical position.

- • *Wheelchair user out of the wheelchair*
  - o The caregiver can assist by merely spotting and/or bringing the wheelchair up the stairs. For the latter, the caregiver proceeds backward up the stairs with the tipped empty wheelchair facing downhill.

- *Variations*
  - The caregiver can carry the wheelchair user "piggy-back" style, with the wheelchair user on the caregiver's back. The wheelchair user holds onto the caregiver with his/her arms over the caregiver's shoulders. The caregiver holds onto the wheelchair user's bent knees.
  - A strong caregiver can carry the wheelchair user "fire-fighter" style with the wheelchair user facing the caregiver and the hips flexed over one of the caregiver's shoulders. The caregiver secures the wheelchair user by wrapping his/her arm around the wheelchair user's knees.
  - Two caregivers can share the load, either front and back or by creating a "seat" of their interlocked hands as described earlier in the "gets from ground into wheelchair" skill.

Special considerations for powered wheelchairs operated by users (Version 3)
- Not applicable.

Special considerations for powered wheelchairs operated by caregivers (Version 4)
- Not applicable.

Special considerations for scooters operated by users (Version 5)
- Not applicable.

7.39 Descends stairs

Versions applicable

Version	Wheelchair Type	Subject Type	Applicable
1	Manual	Wheelchair user	✓
2		Caregiver	✓
3	Powered	Wheelchair user	
4		Caregiver	
5	Scooter	Wheelchair user	

Skill level
- Advanced.

Description
- The wheelchair user and the wheelchair get from the top of a set of stairs to the bottom.

Rationale
- As for the "ascends stairs" skill. Although there is still a potential for injury due to a fall, descent is much less strenuous than ascent. Many wheelchair users who cannot ascend stairs independently can descend them. This skill is not applicable for most powered wheelchairs and scooters.

Prerequisites
- None.

Spotter considerations
- Spotter starting position:
  - If the wheelchair user is proceeding independently down the stairs in the backward direction, the spotter should be behind the wheelchair with the hands near the push-handles.
  - If the wheelchair user is proceeding independently down the stairs in the forward direction, at least two spotters should be involved. One or two spotters should be below the wheelchair with at least one of each spotter's hands near a fixed front part of the wheelchair to resist tipping or runaway. The downhill spotters' other hands may be used to grasp a stair handrail. The uphill spotter should be above the wheelchair with the hands near the push-handles to react to forward or backward tips or runaway.
- Risks requiring spotter intervention:
  - Forward or rear tip or fall.
  - Runaway down the stairs.

Wheelchair Skills Test (WST)

Equipment
- As for the "ascends stairs" skill.
- Because it is often possible to descend stairs that cannot be ascended, an alternative means (e.g., a ramp, lift, test personnel) should be available to get the wheelchair and user to the top of the stairs.
- External aids (e.g., stair lift) may not be used.

Starting positions
- Wheelchair: Facing the top of the stairs, with the leading wheels at least 0.5 m from the edge of the top stair.

Instructions to subject
- Screening questions ("*Can you do it? How?*") are strongly recommended before the subject is allowed to proceed to objective testing.
- "*Get yourself and the wheelchair down the stairs.*"

Capacity criteria
- As for the general scoring criteria, with the clarifications below.
- A "pass with difficulty" should be awarded if:
  o Any effective and safe technique is permitted, as long as at least three stairs are completed.
  o There is excessive jarring as the wheelchair moves from stair to stair.
- A "testing error" score should be awarded if it is not possible to get the wheelchair to the top of the stairs.

Special considerations for manual wheelchairs operated by users (Version 1)
- None.

Special considerations for manual wheelchairs operated by caregivers (Version 2)
- A "pass with difficulty" should be awarded if the caregiver overexerts him/herself.

Special considerations for powered wheelchairs operated by users (Version 3)
- Not applicable.

Special considerations for powered wheelchairs operated by caregivers (Version 4)
- Not applicable.

Special considerations for scooters operated by users (Version 5)
- Not applicable.

Wheelchair Skills Training

General training tips
- As for "ascends stairs."

Special considerations for manual wheelchairs operated by users (Version 1)

- *Out of the wheelchair, on the buttocks or on hands and knees*
  - The procedure is the reverse of the "ascends stairs" skill.

- *In the wheelchair*
  - The safest method is facing up the stairs. The wheelchair user grabs one or both stair rails (Figure 7.111), leans forward enough to keep the casters from lifting off, lowers the rear wheels down one stair, then slides the hands down the rail. The trainer should alert the wheelchair user that this method can be noisy, because the casters and/or footplates bang down each stair; this can be minimized by not leaning too far forward.
  - If the footrests interfere with smooth progression down the stairs and they can be removed, this may be done. The feet are unlikely to be injured as they slide gently from step to step, especially if shoes are worn.

  - *Variations*
    - A variation for the use of two hands on the same rail is for the wheelchair user to turn the trunk toward the rail and reach farther downhill with the rail-side arm (Figure 7.112). This reduces the load on the casters and helps to prevent the wheelchair from turning on the stair.
    - Another option is to face up the stairs as above, but use one hand on the stair handrail and the other hand on the hand-rim of the wheelchair (Figure 7.113). If only a single rail is available, this

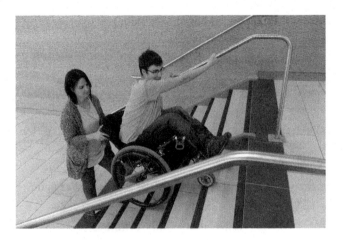

Figure 7.111 Descending stairs in the backward direction, using two hands uphill on the stair rail.

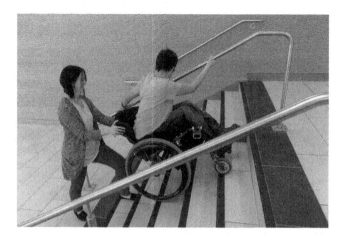

Figure 7.112 Descending stairs in the backward direction, using one hand uphill and one hand downhill on the stair rail.

Figure 7.113 Descending stairs in the backward direction, using one hand uphill on the stair rail and the other hand on the wheelchair hand-rim.

technique can prevent the tendency of the non-rail-side wheel to roll away from the stair riser.

– In the full wheelie position, the wheelchair user can descend forward, one step at a time (Figure 7.114). This is possible if there is an adequate horizontal distance (run) on each step. The wheelchair user drops down one step at a time as for the "descends high curb in wheelie position" skill. The difference is that the casters cannot land after the rear wheels do. The wheelchair user

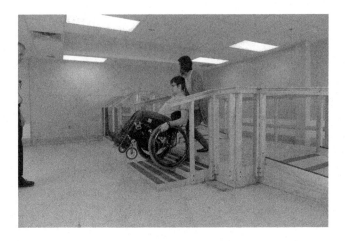

Figure 7.114 Descending low-rise stairs forward in the wheelie position.

instead balances on the rear wheel or, more simply, allows the wheelchair to tilt back after the rear wheels land on the step such that the rear wheels push against the step rise (analogous to the "tilt-rest" skill) before proceeding to the next step. This should be practiced on a single curb first.

- In the full wheelie position, the wheelchair user can descend forward continuously rather than stopping on each step. However, this method is difficult to spot. It is only recommended for a short flight of stairs and when no handrails are available. The wheelchair user approaches the top step as in the previous technique. The difference is that the wheelchair cannot be brought to a stop on each stair. In this technique, the wheelchair user treats the stairs like an incline, with the rear wheels skimming the edges of the steps. If the wheelchair user is going to fall, it is better to fall backward, not forward.
- Using a transient caster pop, the wheelchair user can descend forward continuously rather than stopping on each step. This method is similar to the previous one except that the wheelchair is moving forward as it reaches the edge of the top step. The wheelchair user pops the casters just before the casters reach the drop off. As for the previous technique, the wheelchair user treats the stairs like an incline.
- Descending an escalator is similar to ascending an escalator as described above in the "ascends stairs" skill. The wheelchair user approaches the upper end of the escalator backward, grasps the handrails of the escalator, and allows the wheelchair to be

pulled onto the escalator. While descending the wheelchair user leans forward enough to keep the casters from lifting off the stair. At the bottom, although there is a lip, it usually presents little difficulty because it is first struck by the rear wheels, the large diameters of which allow the relatively unloaded rear wheels to easily roll over.

Special considerations for manual wheelchairs operated by caregivers (Version 2)
- As for the "ascends stairs" skill, but in the reverse direction.

Special considerations for powered wheelchairs operated by users (Version 3)
- Not applicable.

Special considerations for powered wheelchairs operated by caregivers (Version 4)
- Not applicable.

Special considerations for scooters operated by users (Version 5)
- Not applicable.

Chapter 8 Games

In the chapter on individual skills, some variations and activities are described that can be used as means of encouraging varied practice and providing motivation for people learning wheelchair skills. In the current chapter, more detail is provided on some structured games that are suitable for individuals or small groups. Although the importance of organized sports is recognized, descriptions of structured wheelchair sports (e.g., wheelchair basketball, wheelchair rugby, track and field) have not been provided because this is beyond the scope of this book. Depending upon the skill of the participants and the game, spotters may be needed. Note that some of the games or their variations are based around competition and may not be of interest to all participants.

8.1 Line game

Suggested minimum number of players	5
Wheelchair type	• Manual or powered.
Equipment and set-up	• Line grid on floor. Many gyms already have court lines outlined on the floor for participants to follow but, if not, a grid can be easily made using tape.
Instructions	• Participants propel along the lines on the floor. When participants meet each other on a line they must turn around and propel in the opposite direction.
Skills reinforced	• Rolling forward and backward, moving turns, turns in place, spatial awareness.
Variations	• As an ice breaker, have participants introduce themselves when they meet, give each other high fives, shake hands, or wave. • Participants are each given a bingo style sheet with questions in each block such as "brown eyes," or "birthday in April" or "likes to play basketball," etc. When participants meet they must match their new partner with one of the "bingo" blocks and cross it off. The first person to get five blocks in a row wins. • When participants meet, instead of turning around, they propel backward away from their partner until they can turn off down another line, at which point they can propel forward again.

8.2 Traffic lights

Suggested minimum number of players	5
Wheelchair type	• Manual or powered.
Equipment and set-up	• Three colored balloons or signs (green, yellow, and red).
Instructions	• Participants propel wherever they wish in the space provided and at intervals a volunteer or trainer holds up one of the three balloons/signs. Each balloon/sign represents a different instruction. For example, red = stop immediately, yellow = go slowly, and green = go quickly around the room. When the sign is held up, participants must immediately follow the new instructions.
Skills reinforced	• Rolling forward, moving turns (all directions), spatial awareness, and stopping.
Variations	• Trainer shouts out instructions or uses a whistle. • Ask participants to propel backward. • The last person to stop is disqualified. • Use music and encourage participants to go quickly or slowly depending on the speed of the music. When the music stops so must the participants.

8.3 Gears

Suggested minimum number of players	1
Wheelchair type	• Manual or powered.
Equipment and set-up	• Line markings or pylons. • Mark off the room into three different areas or zones.
Instructions	• Participants are instructed to move as slowly as possible through the first area, at a medium speed through the second area, and as quickly as possible in the final area.
Skills reinforced	• Rolling forward, speed control, and braking.
Variations	• The same game, but in the backward direction.

8.4 What time is it, Mr. Wolf?

Suggested minimum number of players	5
Wheelchair type	• Manual.
Equipment and set-up	• None.
Instructions	• The participants are lined up beside each other at the baseline on one side of the room and Mr. Wolf is on the other side of the room, facing away from the participants. • The participants together shout *"What time is it Mr. Wolf?"*. • Mr. Wolf's response corresponds to how many pushes they can give in an attempt to catch the wolf while his/her back is turned. (Example, if Mr. Wolf says that *"It's 3 o'clock"* each participant can move as far as they are able with three pushes.) • If Mr. Wolf says that *"It's dinner time!"* all participants must turn around and propel to the opposite side of the gym without being caught by Mr Wolf.
Skills reinforced	• Rolling forward, stopping, turns in place, and avoiding moving obstacles.
Variations	• See next game (8.5).

8.5 Red light, green light

Suggested minimum number of players	5
Wheelchair type	• Manual or powered.
Equipment and set-up	• None.
Instructions	• Participants line up at one end of the gym. • One participant is chosen as "it" and waits at the opposite end of the gym. • "It" turns his/her back to the rest of the group and calls "*Green light,*" at which point all participants begin to propel forward. • "It" can then call "*Red light*" at any point and turn around quickly. All participants must freeze. • If "it" catches anyone moving when all turns around, that person has to go back to the other end of the gym and start again. • The goal is to tag "it" while his/her back is turned to become the new "it".
Skills reinforced	• Rolling forward, stopping, and turns in place.
Variations	• When "it" calls "*Red light*" and turns around all participants must turn around to face the opposite direction and then freeze. • For more advanced players, when "it" calls "*Red light*" all participants must perform a wheelie. The first person to fall out of the wheelie is "out" (disqualified). The last person "out" becomes the new "it".

8.6 Follow the leader

Suggested minimum number of players	3
Wheelchair type	• Manual or powered.
Equipment and set-up	• None.
Instructions	• A leader is chosen who is responsible for leading the group around the space provided (indoor or outdoor). This leader can perform different skills that the rest of the group tries to copy.
Skills reinforced	• Potential to cover all skill groups depending on the leader.
Variations	• Having more than one group going at once is a good way to divide participants up by skill level.

8.7 Trains

Suggested minimum number of players	6
Wheelchair type	• Manual or powered.
Equipment and set-up	• Flags or equivalent.
Instructions	• Participants form at least two "trains" of three or more people and move around the room with each wheelchair as close as possible to the wheelchair in front of it. Different trainers stand around the room with a flag or other sign. When a flag is raised by a trainer, the trains must propel toward the person holding the flag. The first train to reach the "station" wins that round and the wheelchairs change order.
Skills reinforced	• Rolling forward, speed control, spatial awareness, stopping, moving turns.
Variations	• For manual wheelchairs, the first wheelchair in each train pulls the wheelchairs behind it, with the occupant of the wheelchair behind holding onto the push-handles of the wheelchair in front.

8.8 Slalom

Suggested minimum number of players	1
Wheelchair type	• Manual or powered.
Equipment and set-up	• Start and finish lines about 5 m apart. • Lines and/or walls about 2 m apart on each side to limit how widely the wheelchair can go on either side of the slalom course. • Obstacles—at least four items to turn around, such as pylons, chairs, paper cups, or stones—set up in a line with about 1.2 m between them.
Instructions	• Participant(s) must propel around the obstacles beginning in a prescribed direction (e.g., to the left of the first obstacle) as quickly as possible without touching or displacing the obstacles.
Skills reinforced	• Rolling forward, stopping, spatial awareness, and moving turns.
Variations	• Try different positions for the obstacles, for example, closer together, closer to one wall or line, following a curved path. • Make the course more competitive by counting the number of obstacles displaced and/or measuring the time from the start to the finish line. • Try the course backward. • The same game, but on an incline—up, down, or across.

8.9 Orienteering

Suggested minimum number of players	1
Wheelchair type	• Manual or powered.
Equipment and set-up	• Plan a route outdoors. The route could include obstacles such as different surfaces, cross slopes, curbs, pot holes, inclines, and level changes depending on the skill level of the group. • Photo clue book. • A congratulations sign is placed at each landmark, along with instructions to look at the next photo clue in their book.
Instructions	• In small groups, of participants use photo clues to navigate the route. Each photo shows a landmark that the participants can find (e.g., a tree, a bench, etc.).
Skills reinforced	• Depending on the route used, any combination of skills can be reinforced.
Variations	• When each new clue is found participants can collect objects or cards that could be put together at the end. • Organize different skill level courses so that different routes can be assigned depending on the varying abilities of the groups. • To make this activity more competitive or to be able to assess improvement in ability and skill, performance through the course can be timed.

8.10 Go fish

Suggested minimum number of players	1
Wheelchair type	• Manual or powered.
Equipment and set-up	• Various small objects—pens, magazines, paper clips, coins, coffee cups, peanuts. • Container to hold the objects after they are picked up. • Surfaces on which to place the objects—chairs, tables, shelves, and floor. • Objects are placed on the surfaces around the room.
Instructions	• The participant moves around the room picking up the objects and placing them in the container on his/her lap. When completed, the participant brings the container to the finish point.
Skills reinforced	• Rolling, turning, sideways maneuvering, and reaching.
Variations	• Have participants pick up objects in a certain order. For example, picking up the highest objects first or the lowest objects first. Or only objects of a certain color or shape. • Hide peanuts or similar-sized objects around the room, split participants into teams. The team with the most peanuts at the end of a certain time period wins.

8.11 Circle game

Suggested minimum number of players	7
Wheelchair type	• Manual or powered.
Equipment and set-up	• Space about 5 m². • Participants form a circle, facing the center.
Instructions	• When a participant's name is called by the trainer he or she moves clockwise around the circle until returning to his/her place.
Skills reinforced	• Rolling forward, stopping, spatial awareness, speed control, and moving turns.
Variations	• Cat and mouse: One participant (the cat) propels around the outside of the circle. As he/she does so he/she tags another participant (the mouse) in the circle and the two race in opposite directions around the circle. The last person to return to the original spot is now the "cat".

8.12 Relay race

Suggested minimum number of players	10
Wheelchair type	• Manual or powered.
Equipment and set-up	• Batons (any object will do). • Cones (or equivalent obstacles to turn around).
Instructions	• Divide participants into groups. Each group lines up at one end of the room. When the trainer shouts "*Go,*" the first participant in each group races to the other end of the hall, makes a turn around a cone and returns to his/her group, passing the baton to the next participant. The first group to finish wins.
Skills reinforced	• Rolling forward, moving turns, and reaching.
Variations	• Instead of going around a cone, have participants perform a skill (e.g., wheelie, 360° turn in place, circling the cone twice) once they reach the opposite end of the room. • Instead of returning to the back of the group when a participant finishes the relay, he/she passes the baton to the next person and then follows behind him/her (as in follow-the-leader). This continues until the whole group is led by the final participant around the course. • A series of stations can be spread out at each of which a task must be performed (e.g., picking an object off the floor and placing it on a chair, doing a weight shift for 3 seconds, pouring a cup of water).

8.13 Shrinking space

Suggested minimum number of players	1
Wheelchair type	• Manual or powered.
Equipment and set-up	• Cones (or equivalent obstacles).
Instructions	• A line of cones is placed close to a wall. Each participant attempts to pass between the cones and the wall without touching either. Each time a participant completes this successfully the cones are moved closer to the wall.
Skills reinforced	• Rolling forward or backward, spatial awareness (a good way for participants to learn exactly what gap they can manage in their wheelchairs).
Variations	• Do it backward. • See how quickly participants can get through the space by timing them.

8.14 What's your clearance?

Suggested minimum number of players	1
Wheelchair type	• Manual or powered.
Equipment and set-up	• Objects that can be used to create barriers of increasing widths and heights, for instance by placing them side by side or stacking them. Pieces of wood or bricks are examples. To start the game, a low and narrow obstacle is set up.
Instructions	• Each participant attempts to pass over the obstacle with the object passing between the wheels without the wheels touching the object. Each time a participant completes this successfully the width and/or the height of the obstacle is increased.
Skills reinforced	• Rolling forward or backward, spatial awareness (a good way for participants to learn exactly what clearance they have under their wheelchairs).
Variations	• Sheets of bubble wrap as obstacles are useful to provide audible feedback that a wheel has gone over the obstacle. • Do it backward. • For manual wheelchairs, permit the wheelchair user to use a transient or full wheelie to eliminate the front wheels from consideration. • To add an element of competition, participants can be "out" (disqualified) if they are unable to get over the obstacle without touching it.

8.15 Case open and shut

Suggested minimum number of players	1
Wheelchair type	• Manual or powered.
Equipment and set-up	• A building or structure with different types and styles of doors. • A route description.
Instructions	• Participants are given a route to a series of different doors, returning to the starting point when finished.
Skills reinforced	• Opening and closing a variety of doors.
Variations	• To avoid crowding and delays, teams of 2–3 participants can be routed to the doors in different orders or at intervals (staggered start). • This game can be turned into an orienteering exercise by providing only directions to the next door, where the next set of directions will be posted.

8.16 Stormy seas

Suggested minimum number of players	10
Wheelchair type	• Manual or powered.
Equipment and set-up	• None.
Instructions	• Each participant is given membership to a category of fish (e.g., starfish, shark, octopus). • Participants line up against a wall at one end of the room. • One participant (the fisherman) positions himself/herself in the middle of the room and yells out one of the above categories. When their category is called, the participants must try to get across the room to the other wall without being caught by the fisherman. If tagged, a participant must stop where he/she was caught and he/she becomes seaweed. The seaweed's job is similar to the fisherman's except seaweed cannot move. • If the fisherman yells "*Stormy seas*," all participants try to get to the other side of the room no matter their category.
Skills reinforced	• Rolling forward, moving turns, stopping, and spatial awareness.
Variations	• When a participant is caught he/she becomes an "island" in the sea (rather than seaweed), creating a passive obstacle for the remaining participants to negotiate.

8.17 Simon says

Suggested minimum number of players	3
Wheelchair type	• Manual or powered.
Equipment and set-up	• None.
Instructions	• A leader is chosen who instructs the group to perform certain skills. The participants should only perform the skill when the leader says "*Simon says*" before the instruction. If participant performs a skill when the leader has not said "*Simon says*," that participant is out. The last participant in the game wins.
Skills reinforced	• Potential to cover all skill groups.
Variations	• Simon says mix up: Participants must do the opposite of what "Simon" instructs. For example, if Simon says "*Turn to the right*," participants must turn to the left. If they do what Simon says and not the opposite, they are out.

8.18 Reverse limbo

Suggested minimum number of players	3
Wheelchair type	• Manual or powered.
Equipment and set-up	• Obstacles of various heights.
Instructions	• Participants take turns getting over the obstacle until everyone has completed the task. • The obstacle is then replaced with a higher obstacle. • Participants are eliminated when they can no longer get over the obstacle, and the game continues until only one participant is left.
Skills reinforced	• Getting over obstacles of various heights.
Variations	• Regular limbo: Having an obstacle like a broom handle or rope that can be progressively lowered from an initial position about head high.

8.19 Garbage can basketball

Suggested minimum number of players	6
Wheelchair type	• Manual or powered.
Equipment and set-up	• Ball (any size). • Basket (can be a garbage can or waste basket on a chair).
Instructions	• To begin, create two teams each with an equal number of participants. • Participants are only permitted to carry the basketball for the time it takes them to complete two pushes (if using a manual wheelchair, or equivalent time if using a powered wheelchair) at which point they must either pass the ball to a team member or bounce it on the ground. • Points are scored by getting the ball in the garbage can.
Skills reinforced	• Rolling forward, moving turns, turns in place, spatial awareness, and speed control.
Variations	• Break the game down into its components. Have participants practice bouncing and throwing the ball with a partner. Or practice how to carry the ball for two pushes and then quickly bounce it. Add a quick turn on the end (i.e., push, bounce, fast turn). • Practicing throwing skills by sitting in a circle and passing the ball around. Each time that the ball is passed around the circle without dropping it on the floor, get participants to make the circle bigger by giving one push backward. Then try again. • Practice shooting. Change the height of the net, increasing the height as the participant's skill improves.

8.20 Beach ball chaos

Suggested minimum number of players	3
Wheelchair type	• Manual or powered.
Equipment and set-up	• Beach balls.
Instructions	• A variation on dodge ball, this game can be played as a team or individually. Place one or more balls on the floor and have participants hit the balls with their hand toward members on the other team. Participants must maneuver around balls or block the ball with their hands to avoid the wheelchair getting hit. If the wheelchair is hit by the ball, the participant is frozen until another member of their team high-fives them.
Skills reinforced	• Moving obstacles and reaching.
Variations	• Once a participant is tagged, he/she can be required to come to the back of the court and perform a designated wheelchair skill. • Place two pylons at each end of the room about 2 m apart to create goals. The team that scores the greatest number of goals wins.

8.21 Horse

Suggested minimum number of players	2
Wheelchair type	• Manual or powered.
Equipment and set-up	• None.
Instructions	• Players are numbered and remain in order. Player #1 performs a skill that he/she thinks is possible but that others might find difficult. Starting with Player #2, each player must successfully complete that same skill. If a player is not able to complete the skill, that player is given sequential letters from the word "HORSE." Once a player has all the letters to spell HORSE, he/she is disqualified. Once everyone has tried Player #1's skill, Player #2 presents a different skill and the game continues in the same fashion until there is only one player remaining.
Skills reinforced	• Any combination of skills.
Variations	• Words that are longer (e.g., WHEELCHAIR) or shorter (e.g., PIG) can be used. • Player #1 describes a situation without demonstrating it (e.g., "*Get your wheelchair through the door with your eyes closed*"), to avoid providing clues as to how the skill is accomplished.

Chapter 9 Suggested reading

9.1 References specific to the WSP

The peer-reviewed papers listed below (in alphabetical order) all specifically refer to the WST, WST-Q, and/or the WSTP. A dynamic link on the WSP website (see http://www.wheelchairskillsprogram.ca/eng/publications.php) provides an up-to-the-minute list of such papers and links to them.

Best KL, Kirby RL, Smith C, MacLeod DA. Wheelchair skills training for community-based manual wheelchair users: A randomized controlled trial. *Arch Phys Med Rehabil* 2005;86:2316–23.

Best KL, Kirby RL, Smith C, MacLeod DA. Comparison between performance with a push-rim-activated power-assisted wheelchair and a manual wheelchair on the Wheelchair Skills Test. *Disabil Rehabil* 2006;28:213–20.

Best KL, Miller WC, Huston G, Routhier F, Eng JJ. Pilot study of a peer-led wheelchair training program to improve self-efficacy using a manual wheelchair: A randomized controlled trial. *Arch Phys Med Rehabil* 2016;97:37–44.

Best KL, Miller WC, Routhier F. A description of manual wheelchair skills training curriculum in entry-to-practice occupational and physical therapy programs in Canada. *Disabil Rehabil Assist Technol* 2015;10:401–6.

Best KL, Routhier F, Miller WC. A description of manual wheelchair skills training: Current practices in Canadian rehabilitation centres. *Disabil Rehabil Assist Technol* 2015;10:393–400.

Boucher P, Atrash A, Kelouwani S, Honoré W, Nguyen H, Villemure J, Routhier F, Cohen P, Demers L, Forget R, Pineau J. Design and validation of an intelligent wheelchair towards a clinically-functional outcome. *J Neuroeng Rehabil* 2013;10:58.

Charbonneau R, Kirby RL, Thompson K. Manual wheelchair propulsion by people with hemiplegia: Within-participant comparisons of forward versus backward techniques. *Arch Phys Med Rehabil* 2013;94:1707–13.

Coolen AL, Kirby RL, Landry J, MacPhee AH, Dupuis D, Smith C, Best KL, Mackenzie DE, MacLeod DA. Wheelchair skills training program for clinicians: A randomized controlled trial with occupational therapy students. *Arch Phys Med Rehabil* 2004;85:1160–7.

Eriks-Hoogland I, de Groot S, Snoek G, Stucki G, Post M, van der Woude L. Association of shoulder problems in persons with spinal cord injury at discharge from inpatient rehabilitation with activities and participation 5 years later. *Arch Phys Med Rehabil* 2016;97:84–91.

Fliess-Douer O, Vanlandewijck YC, Lubel Manor G, Van Der Woude LH. A systematic review of wheelchair skills tests for manual wheelchair users with a spinal cord injury: Towards a standardized outcome measure. *Clin Rehabil* 2010;24:867–86.

Furmaniuk L, Cywińska-Wasilewska G, Kaczmarek D. Influence of long-term wheelchair rugby training on the functional abilities in persons with tetraplegia over a two-year post-spinal cord injury. *J Rehabil Med* 2010;42:688–90.

Giesbrecht EM, Miller WC, Eng JJ, Mitchell IM, Woodgate RL, Goldsmith CH. Feasibility of the enhancing participation in the community by improving wheelchair skills (EPIC Wheels) program: Study protocol for a randomized controlled trial. *Trials* 2013;14:350.

Giesbrecht EM, Miller WC, Mitchell IM, Woodgate RL. Development of a wheelchair skills home program for older adults using a participatory action design approach. *Biomed Res Int* 2014;2014:172434.

Giesbrecht EM, Wilson N, Schneider A, Bains D, Hall J, Miller WC. Preliminary evidence to support a "boot camp" approach to wheelchair skills training for clinicians. *Arch Phys Med Rehabil* 2015;96:1158–61.

Hosseini SM, Oyster ML, Kirby RL, Harrington AL, Boninger ML. Manual wheelchair skills capacity predicts quality of life and community integration in persons with spinal cord injury. *Arch Phys Med Rehabil* 2012;93:2237–43.

Inkpen P, Parker K, Kirby RL. Manual wheelchair-skills capacity versus performance. *Arch Phys Med Rehabil* 2012;93:1009–13.

Jung HS, Park G, Kim YS, Jung HS. Development and evaluation of one-hand drivable manual wheelchair device for hemiplegic patients. *Appl Ergon* 2015;48:11–21.

Kilkens OJ, Post MW, Dallmaijer AJ, Seelen HA, van der Woude LH. Wheelchair skills tests: A systematic review. *Clin Rehabil* 2003;17:418–30.

Kirby RL, Adams CD, MacPhee AH, Coolen AL, Harrison ER, Eskes GA, Smith C, MacLeod DA, Dupuis DJ. Wheelchair-skill performance: Controlled comparison between people with hemiplegia and able-bodied people simulating hemiplegia. *Arch Phys Med Rehabil* 2005;86:387–93.

Kirby RL, Corkum CG, Smith C, Rushton P, MacLeod DA, Webber A. Comparing performance of manual wheelchair skills using new and conventional rear anti-tip devices: Randomized controlled trial. *Arch Phys Med Rehabil* 2008;89:480–5.

Kirby RL, Crawford KA, Smith C, Thompson KJ, Sargeant JM. A wheelchair workshop for medical students improves knowledge and skills: A randomized controlled trial. *Am J Phys Med Rehabil* 2011;90:197–206.

Kirby RL, Dupuis DJ, MacPhee AH, Coolen AL, Smith C, Best KL, Newton AM, Mountain AD, MacLeod DA, Bonaparte JP. The Wheelchair Skills Test (version 2.4): Measurement properties. *Arch Phys Med Rehabil* 2004;85:794–804.

Kirby RL, MacDonald B, Smith C, MacLeod DA, Webber A. Comparison between a tilt-in-space wheelchair and a manual wheelchair equipped with a new rear anti-tip device from the perspective of the caregiver. *Arch Phys Med Rehabil* 2008;89:1811–5.

Kirby RL, Mifflen NJ, Thibault DL, Smith C, Best KL, Thompson KJ, MacLeod DA. The manual wheelchair-handling skills of caregivers and the effect of training. *Arch Phys Med Rehabil* 2004;85:2011–9.

Kirby RL, Miller WC, Routhier F, Demers L, Mihailidis A, Polgar JM, Rushton PW, Titus L, Smith C, McAllister M, Theriault C, Thompson K, Sawatzky B. Effectiveness of a wheelchair skills training program for powered wheelchair users: A randomized controlled trial. *Arch Phys Med Rehabil* 2015;96:2017–26.e3.

Kirby RL, Swuste J, Dupuis DJ, MacLeod DA, Monroe R. The Wheelchair Skills Test: A pilot study of a new outcome measure. *Arch Phys Med Rehabil* 2002;83:10–8.

Kirby RL, Walker R, Smith C, Best K, MacLeod DA, Thompson K. Manual wheelchair-handling skills by caregivers using new and conventional rear anti-tip devices: A randomized controlled trial. *Arch Phys Med Rehabil* 2009;90:1680–4.

Lemay V, Routhier F, Noreau L, Phang SH, Ginis KA. Relationships between wheelchair skills, wheelchair mobility and level of injury in individuals with spinal cord injury. *Spinal Cord* 2012;50:37–41.

Lindquist NJ, Loudon PE, Magis TF, Rispin JE, Kirby RL, Manns PJ. Reliability of the performance and safety scores of the Wheelchair Skills Test version 4.1 for manual wheelchair users. *Arch Phys Med Rehabil* 2010;91:1752–7.

MacPhee AH, Kirby RL, Coolen AL, Smith C, MacLeod DA, Dupuis DJ. Wheelchair skills training program: A randomized clinical trial of wheelchair users undergoing initial rehabilitation. *Arch Phys Med Rehabil* 2004;85:41–50.

Morgan KA, Engsberg JR, Gray DB. Important wheelchair skills for new manual wheelchair users: Health care professional and wheelchair user perspectives. *Disabil Rehabil Assist Technol* 2015;3;1–11.

Morgan KA, Tucker SM, Klaesner JW, Engsberg JR. A motor learning approach to training wheelchair propulsion biomechanics for new manual wheelchair users: A pilot study. *J Spinal Cord Med* 2015; December 16 [Epub ahead of print].

Mortenson WB, Demers L, Rushton PW, Auger C, Routhier F, Miller WC. Exploratory validation of a multidimensional power wheelchair outcomes toolkit. *Arch Phys Med Rehabil* 2015;96:2184–93.

Mortenson WB, Miller WC, Polgar JM. Measurement properties of the Late Life Disability Index among individuals who use power wheelchairs as their primary means of mobility. *Arch Phys Med Rehabil* 2014;95:1918–24.

Mountain AD, Kirby RL, Eskes GA, Smith C, Duncan H, MacLeod DA, Thompson K. Ability of people with stroke to learn powered wheelchair skills: A pilot study. *Arch Phys Med Rehabil* 2010;91:596–601.

Mountain AD, Kirby RL, Smith C. The Wheelchair Skills Test, version 2.4: Validity of an algorithm-based questionnaire version. *Arch Phys Med Rehabil* 2004;85:416–23.

Mountain AD, Kirby RL, Smith C, Eskes G, Thompson K. Powered wheelchair skills training for persons with stroke: A randomized controlled trial. *Am J Phys Med Rehabil* 2014;93:1031–43.

Nagy J, Winslow A, Brown JM, Adams L, O'Brien K, Boninger M, Nemunaitis G. Pushrim kinetics during advanced wheelchair skills in manual wheelchair users with spinal cord injury. *Top Spinal Cord Inj Rehabil* 2012;18:140–2.

Nelson AL, Groer S, Palacios P, Mitchell D, Sabharwal S, Kirby RL, Gavin-Dreschnack D, Powell-Cope G. Wheelchair-related falls in veterans with spinal cord injury residing in the community: A prospective cohort study. *Arch Phys Med Rehabil* 2010;91:1166–73.

Newton AM, Kirby RL, MacPhee AH, Dupuis DJ, MacLeod DA. Evaluation of manual wheelchair skills: Is objective testing necessary or would subjective estimates suffice? *Arch Phys Med Rehabil* 2002;83:1295–9.

Oztürk A, Ucsular FD. Effectiveness of a wheelchair skills training programme for community-living users of manual wheelchairs in Turkey: A randomized controlled trial. *Clin Rehabil* 2011;25:416–24.

Phang SH, Martin Ginis KA, Routhier F, Lemay V. The role of self-efficacy in the wheelchair skills-physical activity relationship among manual wheelchair users with spinal cord injury. *Disabil Rehabil* 2012;34:625–32.

Pradon D, Pinsault N, Zory R, Routhier F. Could mobility performance measures be used to evaluate wheelchair skills? *J Rehabil Med* 2012;44:276–9.

Routhier F, Kirby RL, Demers L, Depa M, Thompson K. Efficacy and retention of the French-Canadian version of the wheelchair skills training program for manual wheelchair users: A randomized controlled trial. *Arch Phys Med Rehabil* 2012;93:940–8.

Rushton PW, Kirby RL, Miller WC. Manual wheelchair skills: Objective testing versus subjective questionnaire. *Arch Phys Med Rehabil* 2012;93:2313–8.

Rushton PW, Kirby RL, Routhier F, Smith C. Measurement properties of the Wheelchair Skills Test—Questionnaire for powered wheelchair users. *Disabil Rehabil Assist Technol* 2014;20:1–7.

Sakakibara BM, Miller WC. Prevalence of low mobility and self-management self-efficacy in manual wheelchair users and the association with wheelchair skills. *Arch Phys Med Rehabil* 2015;96:1360–3.

Sakakibara BM, Miller WC, Eng JJ, Backman CL, Routhier F. Preliminary examination of the relation between participation and confidence in older manual wheelchair users. *Arch Phys Med Rehabil* 2013;94:791–4.

Sakakibara BM, Miller WC, Eng JJ, Backman CL, Routhier F. Influences of wheelchair-related efficacy on life-space mobility in adults who use a wheelchair and live in the community. *Phys Ther* 2014;94:1604–13.

Sakakibara BM, Miller WC, Routhier F, Backman CL, Eng JJ. Association between self-efficacy and participation in community-dwelling manual wheelchair users aged 50 years or older. *Phys Ther* 2014;94:664–74.

Sakakibara BM, Miller WC, Souza M, Nikolova V, Best KL. Wheelchair skills training to improve confidence with using a manual wheelchair among older adults: A pilot study. *Arch Phys Med Rehabil* 2013;94:1031–7.

Sawatzky B, DiGiovine C, Berner T, Roesler T, Katte L. The need for updated clinical practice guidelines for preservation of upper extremities in manual wheelchair users: A position paper. *Am J Phys Med Rehabil* 2015;94:313–24.

Sawatzky B, Hers N, MacGillivray MK. Relationships between wheeling parameters and wheelchair skills in adults and children with SCI. *Spinal Cord* 2015;53:561–4.

Smith C, Kirby RL. Manual wheelchair skills capacity and safety of residents of a long-term-care facility. *Arch Phys Med Rehabil* 2011;92:663–9.

Sorrento GU, Archambault PS, Routhier F, Dessureault D, Boissy P. Assessment of joystick control during the performance of powered wheelchair driving tasks. *J Neuroeng Rehabil* 2011;8:31.

Tangsagulwatthana S, Sawattikano N, Kovindha A. Wheelchair skills training for individuals with spinal cord injury: A pilot study. *Thai J Phys Ther* 2010;32:173–80.

Toro ML, Eke C, Pearlman J. The impact of the World Health Organization 8-steps in wheelchair service provision in wheelchair users in a less resourced setting: A cohort study in Indonesia. *BMC Health Serv Res* 2016;16:26.

9.2 Other relevant reading materials

Agree EM. The influence of personal care and assistive devices on the measurement of disability. *Soc Sci Med* 1999;48:427–43.

Agree EM, Freedman VA. Incorporating assistive devices into community-based long-term care: An analysis of the potential for substitution and supplementation. *J Aging Health* 2000;12:426–50.

Archambault PS, Sorrento G, Routhier F, Boissy P. Assessing improvement of powered wheelchair driving skills using a data logging system. In: *Proceedings of the RESNA Annual Conference*, Las Vegas, Nevada, June 28–29, 2010.

Armstrong W, Reisinger KD, Smith WK. Evaluation of CIR-whirlwind wheelchair and service provision in Afghanistan. *Disabil Rehabil* 2007;29:935–48.

Askari S, Kirby RL, Parker K, Thompson K, O'Neill J. Wheelchair Propulsion Test: Development and measurement properties of a new test for manual wheelchair users. *Arch Phys Med Rehabil* 2013;94:1690–8.

Auger C, Demers L, Gelinas I, Jutai J, Fuhrer MJ, DeRuyter F. Powered mobility for middle-aged and older adults: Systematic review of outcomes and appraisal of published evidence. *Am J Phys Med Rehabil* 2008;87:666–80.

Auger C, Demers L, Gelinas I, Miller WC, Jutai JW, Noreau L. Life-space mobility of middle-aged and older adults at various stages of usage of power mobility devices. *Arch Phys Med Rehabil* 2010;91:765–73.

Auger C, Demers L, Gélinas I, Routhier F, Mortenson WB, Miller WC. Reliability and validity of the telephone administration of the wheelchair outcome measure (WhOM) for middle-aged and older users of power mobility devices. *J Rehabil Med* 2010;42:574–81.

Axelson PW, Minkel J, Perr A, Hubbard B, Butler C. *The Manual Wheelchair Training Guide.* 2nd Edition. PAX Press, Minden, NV, 2013.

Axelson PW, Minkel J, Perr A, Yamada D. *The Powered Wheelchair Training Guide.* 2nd Edition. PAX Press, Minden, NV, 2013.

Baker PS, Bodner EV, Allman RM. Measuring life-space mobility in community-dwelling older adults. *J Am Geriatr Soc* 2003;51:1610–4.

Barker DJ, Reid D, Cott C. The experience of senior stroke survivors: Factors in community participation among wheelchair users. *Can J Occup Ther* 2006;73:18–25.

Barker-Collo S, Feigin V. The impact of neuropsychological deficits on functional stroke outcomes. *Neuropsychol Rev* 2006;16,53–64.

Berg K, Hines M, Allen S. Wheelchair users at home: Few home modifications and many injurious falls. *Am J Public Health* 2002;92:48.

Borg J, Larsson S, Ostergren PO, Rahman AS, Bari N, Khan AH. User involvement in service delivery predicts outcomes of assistive technology use: A cross-sectional study in Bangladesh. *BMC Health Serv Res* 2012;12:330.

Brandt A, Iwarsson S, Stahle A. Older people's use of powered wheelchairs for activity and participation. *J Rehabil Med* 2004;36:70–7.

Bullard S, Miller SE. Comparison of teaching methods to learn a tilt and balance wheelchair skill. *Precept Mot Skills* 2001;93:131–8.

Calder CJ, Kirby RL. Fatal wheelchair-related accidents in the USA. *Am J Phys Med Rehabil* 1990;69:184–90.

Clarke P, Colantonio A. Wheelchair use among community-dwelling older adults: Prevalence and risk factors in a national sample. *Can J Aging* 2005;24:191–8.

Cowan RE, Boninger ML, Sawatzky BJ, Mazoyer BD, Cooper RA. Preliminary outcomes of the SmartWheel Users' Group database: A proposal framework for clinicians to objectively evaluate manual wheelchair propulsion. *Arch Phys Med Rehabil* 2008;89:260–8.

Cullen B, O'Neill B, Evans JJ. Neuropsychological predictors of powered wheelchair use: A prospective follow-up study. *Clin Rehabil* 2008;22:836–46.

Davies A, De Souza LH, Frank AO. Changes in the quality of life in severely disabled people following provision of powered indoor/outdoor chairs. *Disabil Rehabil* 2003;25:286–90.

Demers L, Fuhrer MJ, Jutai JW, Scherer MJ, Pervieux I, DeRuyter F. Tracking mobility-related assistive technology in an outcomes study. *Assist Technol* 2008;20:73–83.

Denison I. *Manual Wheelchair Skills. Guidelines for Instructing.* http://www.assistive-technology.ca/docs/mwcskillsg.pdf. Accessed March 8, 2016.

Devos H, Akinwuntan AE, Nieuwboer A, Truijen S, Tant M, De Weerdt W. Screening for fitness to drive after stroke: A systematic review and meta-analysis. *Neurology* 2011;76:747–56.

Dijcks BPJ, De Witte LP, Gelderblom GJ, Wessels RD, Soede M. Non-use of assistive technology in The Netherlands: A non-issue? *Disabil Rehabil Assist Technol* 2006;1:97–102.

Donnelly C, Eng JJ, Hall J, Alford L, Giachino R, Norton K, Kerr DS. Client-centered assessment and the identification of meaningful treatment goals for individuals with a spinal cord injury. *Spinal Cord* 2004;42:302–7.

Edwards K, McCluskey A. A survey of adult power wheelchair and scooter users. *Disabil Rehabil Assist Technol* 2010;5:411–9.

Evans R. The effect of electrically powered indoor/outdoor wheelchairs on occupation: A study of users' views. *Brit J Occup Ther* 2000;63:547–53.

Flagg JF. Wheeled mobility demographics. In: S Bauer and ME Buning (Eds), *Industry Profile on Wheeled Mobility*. Chapter 1. RERC on Technology Transfer, State University of New York, Buffalo, NY, February 2009, pp. 7–29.

Fliess-Douer O, Vanlandewijck YC, van der Woude LH. Most essential wheeled mobility skills for daily life: An international survey among paralympic wheelchair athletes with spinal cord injury. *Arch Phys Med Rehabil* 2012;93:629–35.

Frank A, Neophylou C, Frank J, Souza L. Electric-powered indoor/outdoor wheelchairs (EPIOCs): Users's views of influence on family, friends and carers. *Disabil Rehabil Assist Technol* 2010;5:327–38.

Frank AO, Ward J, Orwell NJ, McCullagh C, Belcher M. Introduction of a new NHS electric powered indoor/outdoor chair (EPIOC) service: Benefits, risks and implications for pre-scribers. *Clin Rehabil* 2000;14:665–73.

Gaal RP, Rebholtz N, Hotchkiss RD, Pfaelzer PF. Wheelchair rider injuries: Causes and conse-quences for wheelchair design and selection. *J Rehabil Res Dev* 1997;34:58–71.

Greer N, Brasure M, Wilt TJ. Wheeled mobility (wheelchair) service delivery: Scope of the evidence. *Ann Intern Med* 2012;156:141–6.

Guerette P, Tefft D, Furumasu J. Pediatric powered wheelchairs: Results of a national survey of providers. *Assist Technol* 2005;17:144–58.

Häggblom-Kronlöf G, Sonn U. Use of assistive devices—A reality full of contradictions in elderly persons' everyday life. *Disabil Rehabil Assist Technol* 2007;2:335–45.

Hall K, Partoy J, Tenenbaum S, Dawson DR. Power mobility driving training for seniors: A pilot study. *Assist Technol* 2005;17:47–56.

Heinemann AW, Dijkers MP, Ni P, Tulsky DS, Jette A. Measurement properties of the Spinal Cord Injury-Functional Index (SCI-FI) short forms. *Arch Phys Med Rehabil* 2014;95:1289–97.

Hocking C. Function or feeling: Factors in abandonment in assistive devices. *Technol Disabil* 1999;11:3–11.

Hoenig H, Landerman LR, Shipp KM, Pieper C, Pieper C, Richardson M, Pahel N, George L. A clinical trial of rehabilitation expert clinician versus usual care for providing manual wheelchairs. *J Am Geriatr Soc* 2005;53:1712–20.

Hoenig H, Pieper C, Branch LG, Cohen HJ. Effect of motorized scooters on physical performance and mobility: A randomized clinical trial. *Arch Phys Med Rehabil* 2007;88:279–86.

Hoenig H, Pieper C, Zolkewitz M, Schenkman M, Branch LG. Wheelchair users are not neces-sarily wheelchair bound. *J Am Geriatr Soc* 2002;50:645–54.

Hubbard SL, Fitzgerald SG, Reker M, Boninger ML, Cooper RA, Kazis LE. Demographic characteristics of veterans who received wheelchairs and scooters from Veterans Health Administration. *J Rehabil Res Dev* 2006;43:831–44.

Hubbard SL, Fitzgerald SG, Vogel B, Reker DM, Cooper RA, Boninger ML. Distribution and cost of wheelchairs and scooters provided by Veterans Health Administration. *J Rehabil Res Dev* 2007;44:581–92.

Jannink MJ, Erren-Wolters CV, de Kort AC, van der Kooij H. An electric scooter simulation program for training the driving skills of stroke patients with mobility problems: A pilot study. *Cyberpsychol Behav* 2008;11:751–4.

Kangas KM. Clinical assessment and training strategies for the child's mastery of independent powered mobility. In: J Furamasu (Ed), *Pediatric Powered Mobility: Developmental Perspectives, Technical Issues, Clinical Approaches*. RESNA Press, Arlington, Virginia, 1997.

Karmarkar AM, Collins DM, Wichman T, Franklin A, Fitzgerald SG, Dicianno BE, Pasquina PF, Cooper RA. Prosthesis and wheelchair use in veterans with lower-limb amputation. *J Rehabil Res Dev* 2009;46:567–76.

Kilkens OJE, Post MWM, Dallmeijer AJ, van Asbeck FWA, van der Woude LHV. Relationship between manual wheelchair skill performance and participation of persons with spinal cord injuries 1 year after discharge from inpatient rehabilitation. *J Rehabil Res Dev* 2005;42(3 Supp 1):65–73.

King EC, Dutta T, Gorski SM, Holliday PJ, Fernie GR. Design of built environments to accommodate mobility scooter users: Part II. *Disabil Rehabil Assist Technol* 2011;6:432–9.

Kirby RL, Ackroyd-Stolarz SA, Brown MG, Kirkland SA. Wheelchair-related accidents caused by tips and falls among noninstitutionalized users of manually propelled wheelchairs in Nova Scotia. *Am J Phys Med Rehabil* 1994;73:319–30.

Kittel A, Di Marco A, Stewart H. Factors influencing the decision to abandon manual wheelchairs for three individuals with a spinal cord injury. *Disabil Rehabil* 2002;24;106–14.

Krause J, Carter RE, Brotherton S. Association of mode of locomotion and independence in locomotion with long-term outcomes after spinal cord injury. *J Spinal Cord Med* 2009;32:237–48.

Leduc BE, Lepage Y. Health-related quality of life after SCI. *Disabil Rehabil* 2002;24:196–202.

Lofqvist C, Pettersson C, Iwarsson S, Brandt A. Mobility and mobility-related participation outcomes of powered wheelchair and scooter interventions after 4-months and 1-year use. *Disabil Rehabil Assist Technol* 2012;7:211–8.

Magill RA. *Motor Learning and Control: Concepts and Applications.* 9th Edition. McGraw-Hill, New York, 2011.

McClure LA, Boninger ML, Oyster ML, Williams S, Houlihan B, Lieberman JA, Cooper RA. Wheelchair repairs, breakdown, and adverse consequences for people with traumatic spinal cord injury. *Arch Phys Med Rehabil* 2009;90:2034–8.

Mortenson WB, Clarke LH, Best K. Prescribers experiences with powered mobility prescription among older adults. *Am J Occup Ther* 2013;67:100–7.

Mortenson WB, Demers L, Fuhrer MJ, Jutai J, Lenker J, DeRuyter F. Effects of an assistive technology intervention on older adults with disabilities and their informal caregivers: An exploratory randomized control trial. *Am J Phys Med Rehabil* 2013; 92:297–306.

Mortenson WB, Miller WC. The wheelchair procurement process: Perspectives of clients and prescribers. *Can J Occup Ther* 2008;75:177–85.

Mortenson WB, Miller WC, Backman CL, Oliffe JL. Predictors of mobility among wheelchair using residents in long-term care. *Arch Phys Med Rehabil* 2011;92:1587–93.

Mortenson WB, Miller WC, Backman CL, Oliffe JL. Association between mobility, participation and wheelchair related factors in residents who use wheelchairs as their primary means of mobility. *J Am Geriatr Soc* 2012;60:1310–5.

Mortenson WB, Miller WC, Boily J, Steele B, Odell L, Crawford EM, Desharnais G. Perceptions of power mobility use and safety within residential facilities. *Can J Occup Ther* 2005;72:142–52.

Mortenson WB, Miller WC, Miller-Polgar J. Measuring wheelchair intervention outcomes: Development of the wheelchair outcome measure. *Disabil Rehabil Assist Technol* 2007;2:275–85.

Nitz JC. Evidence from a cohort of able bodied adults to support the need for driver training for motorized scooters before community participation. *Patient Educ Couns* 2008;70:276–80.

Niv A, Weiss P, Ratzon N. The effectiveness of combining occupational therapy intervention with computerized training for improved driving on the electric scooter. *Isr J Occup Ther* 2009;18:E14–5.

Noreau L, Fougeyrollas P. Long-term consequences of spinal cord injury on social participation: The occurrence of handicap situations. *Disabil Rehabil* 2000;22:170–80.

Pettersson I, Ahlström G, Törnquist K. The value of an outdoor powered wheelchair with regard to the quality of life of persons with stroke: A follow-up study. *Assist Technol* 2007;19:143–53.

Rousseau-Harrison K, Rochette A, Routhier F, Dessureault D, Thibault F, Cote O. Impact of wheelchair acquisition on social participation. *Disabil Rehabil Assist Technol* 2009;4:344–52.

Rudman DL, Hebert D, Reid D. Living in a restricted occupational world: The occupational experiences of stroke survivors who are wheelchair users and their caregivers. *Can J Occup Ther* 2006;73:141–52.

Rushton PW, Miller WC, Kirby RL, Eng JJ. Measure for the assessment of confidence with manual wheelchair use (WheelCon-M) Version 2.1: Reliability and validity. *J Rehabil Med* 2013;45:61–7.

Rushton PW, Miller WC, Kirby RL, Eng JJ, Yip J. Development and content validation of the Wheelchair Use Confidence Scale: A mixed-methods study. *Disabil Rehabil Assist Technol* 2011;6:57–66.

Salminen A-L, Brandt A, Sameulsson K, Toytari O, Malmivaara A. Mobility devices to promote activity and participation: A systematic review. *J Rehabil Med* 2009;41:697–706.

Sawatzky B, Rushton PW, Denison I, McDonald R. Wheelchair skills training programme for children: A pilot study. *Aust Occup Ther J* 2012;59:2–9.

Smith E, Sakakibara BM, Miller WC. A review of factors influencing participation in social and community activities for wheelchair users. *Disabil Rehabil Assist Technol* 2016;11:361–74.

Taylor DH, Hoenig H. The effect of equipment usage and residual task difficulty on use of personal assistance, days in bed, and nursing home placement. *J Am Geriatr Soc* 2004;52:72–9.

Taylor-Schroeder S, LaBarbera J, McDowell S, Zanca JM, Natale A, Mumma S, Gassaway J, Backus D. The SCI Rehab Project: Treatment time spent in SCI rehabilitation. Physical therapy treatment time during inpatient spinal cord injury rehabilitation. *J Spinal Cord Med* 2011;34:149–61.

Ummat S, Kirby RL. Nonfatal wheelchair-related accidents reported to the National Electronic Injury Surveillance System. *Am J Phys Med Rehabil* 1994;73:163–7.

van Zeltzen JM, de Groot S, Post MWM, Slootman JR, van Bennekom CAM, van der Woude LHV. Return to work after spinal cord injury. *Am J Phys Med Rehabil* 2009;88:47–56.

Wessels R, Kijcks B, Soede M, Gelderblom GJ, De Witte L. Non-use of provided assistive technology devices: A literature overview. *Technol Disabil* 2003;15:231–8.

World Health Organization (WHO). *Guidelines on the Provision of Wheelchairs in Less-Resourced Settings.* www.who.int/disabilities/publications/technology/wheelchairguidelines/en/. 2008. Accessed March 8, 2016.

World Health Organization. *International Classification of Functioning, Disability and Health (ICF).* http://www.who.int/classifications/icf/en/. Accessed March 8, 2016.

Worobey L, Oyster M, Nemunaitis G, Cooper R, Boninger ML. Increases in wheelchair breakdowns, repairs, and adverse consequences for people with traumatic spinal cord injury. *Am J Phys Med Rehabil* 2012;91:463–9.

Xiang H, Chany A-M, Smith GA. Wheelchair related injuries treated in US emergency departments. *Inj Prev* 2006;12:8–11.

Zanca JM, Dijkers MP, Hsieh CH, Heinemann AW, Horn SD, Smout RJ, Backus D. Group therapy utilization in inpatient spinal cord injury rehabilitation. *Arch Phys Med Rehabil* 2013;94(4 Suppl):S145–53.

Zanka JM, Natale A, LaBarbera J, Schroeder ST, Gassaway J, Backus D. Group physical therapy during inpatient rehabilitation for acute spinal cord injury: Findings from the SCI Rehab Study. *Phys Ther* 2012;91:1877–91.

Appendix 1: Lesson plans

Before each WSTP session, the trainer should have a plan for how each session will be conducted as well as a plan for the series of sessions. The lesson plans will be affected by whether the training will be one-on-one or in a group, by the group size, group makeup (diagnoses accounting for wheelchair use, skill level), session specifics (e.g., the number, frequency, and duration of sessions), training facilities, and by the number of trainers and spotters available.

The sample lesson plans below are general templates for one-on-one training in sessions scheduled for 30 minutes following an intake session of 40 minutes.

Intake session (40 minutes)*

A. Welcome (2 minutes).
   - Explain purpose of this and subsequent sessions.
   - Obtain informed consent to proceed.
B. Perform an intake assessment (25 minutes).
   - Document demographic, clinical, and wheelchair-experience data.
   - Identify any contraindications for testing or training.
   - Document wheelchair specifications.
   - Wheelchair skills assessment (WST-Q and/or WST).
C. Goal setting (5 minutes).
   - From the intake assessment and discussion with the learner, identify and record a set of relevant and potentially achievable training goals.
D. Begin training (5 minutes).
   - Begin work on an initial goal so that the learner goes away with at least one skill to practice before the next session.
E. Closing (3 minutes).
   - Describe the nature of subsequent sessions.
   - Schedule the next session.

---

* Times are rough guidelines only.

339

- Assign homework.
- Answer any questions that the learner may have.
- Provide strong encouragement.
- Complete any final documentation of the session.

Subsequent sessions (30 minutes)

A. Welcome and warm-up (5 minutes).
- Check status: Any new health concerns since the last session? Any after-effects from the last session? Any practice since the last session? Any wheelchair changes?
- Review the goals and planned activities for the current session.
- Questions and answers.
- Warm-up activity.
B. Practice skills that have already been acquired but that need work (10 minutes).
- Random order, but begin with less stressful ones until the learner is warmed up.
- Variety of settings.
- Trainer role: Provide structure, safety, and minimal feedback.
- This portion of the session can also serve to provide conditioning, if the sessions are scheduled often enough to serve in that capacity (i.e., at least three times a week).
- Games can be a fun way to carry out this stage of the session.
C. Practice a skill that has not been acquired yet (10 minutes).
- Trainer role: Provide structure, safety, instructions, demonstration, and feedback.
D. Closing (5 minutes).
- Questions and answers.
- Plan next session content.
- Assign homework.
- Schedule next session.
- Complete any final documentation of the session.

Appendix 2: Sample outlines for group training sessions

A2.1 Small groups

This sample outline is intended to assist trainers working with groups of 5–10 participants, all of whom use two-hand wheelchair propulsion. The outline has been structured to be used over a period of 6 sessions, each lasting 1 hour. A general template begins on the next page.

The content should be adjusted depending upon the skill level of the participants and the local setting. This sample outline is intended to be used in combination with other materials found earlier in this book and on the WSP website.

Prior to any training sessions, for the purposes of this example, it is assumed that each participant has already been seen individually to perform an intake assessment that includes the following elements:

- Record contact information (phone numbers, email address, and next-of-kin).
- Document demographic, clinical, and wheelchair-experience data.
- Identify any contraindications for training.
- Document wheelchair specifications.
- Wheelchair skills assessment (WST-Q and/or WST).
- Identify and record a set of relevant and potentially achievable training goals.

It is also possible to obtain most of this intake information in a group setting. However, if the WST is to be used as an outcome measure, the WST should not be performed as a group because witnessing others may affect the participant's technique.

General session template that applies to all sessions unless otherwise specified

ADVANCE PREPARATION BY THE TRAINER

- Confirm that the space has been booked, if necessary.
- Ensure that participants and training personnel know the session date, time, and location. (It is a good idea to remind participants and training personnel of the upcoming session if it will be more than a week since the previous session.)
- Obtain, prepare, or review materials needed for every session:
  o List of participants and their contact information.
  o Attendance sheet.
  o Documentation of intake data for each participant.
  o Clip-board and pen or pencil for each participant.
  o Pencil sharpener.
  o List of goals for each participant.
  o Name tags for participants and training personnel.
  o Whistle or other noise maker for calling attention.
  o Air pump for tires.
  o Tool kit for urgent repairs or adjustments.
- Obtain, prepare, or review any materials needed for this specific session:
  o Review the appropriate sections of this book or the WSP Manual.
  o Review on-line videos of the skills to be covered.
  o If the session being prepared for is the first session:
    − Signage directing participants to the arrival area.
  o If the session being prepared for is the final session:
    − Evaluation forms for the participants to complete.
    − Report cards.
    − Certificates.
    − Humorous prizes.

THE ACTUAL TRAINING SESSION

- Arrival of participants (5 minutes).
  o Greet participants as they arrive.
  o If this is the first session:
    − Direct participants to where they should hang up their coats and knapsacks.
    − Let them know that an air pump and tools are available for urgent maintenance.

- Each participant picks up his/her name tag, clipboard with goal list attached, and pen or pencil.
- Session opening (10 minutes).
  - Call to order: Form a circle ("huddle").
  - Introductions (mostly at first session but at subsequent sessions if there are any new people in attendance).
  - Record attendance.
  - At first session, achieve consensus on rules and post them on the wall for reference purposes:
    - Attend all sessions (notify trainer if unable to attend for some reason—provide trainer contact information).
    - Be on time for sessions.
    - Turn off cell phones during sessions.
    - During huddles or explanations, only one person taking at a time.
    - Do not attempt any skill that you are not sure that you can do safely without a spotter.
    - Agree on a penalty for rule violation (e.g., sing a song, do a dance).
  - Check for any participant status changes since the last session (the intake session, in the case of the first session):
    - Any after-effects from the last training session?
    - Any practice carried out (encourage)?
    - Any wheelchair changes?
    - Any new health concerns? (Invite participants to speak to you privately, before beginning the warm-up activity.)
  - Review the general purpose of the training sessions—to improve specific wheelchair skills in order to prevent injury and overcome environmental barriers.
  - Have each participant independently review his/her overall goals for the series of training sessions, revising them if appropriate.
  - Explain the planned activities for the current session.
  - Warm-up activity:
    - An activity designed to warm up the muscles, heart, and lungs.
    - The activity should include some skills that have already been learned or, if the first session, skills that the trainer knows from the intake assessments that all participants can perform.
    - A game (see Games section of this book or the WSP Manual) can be a fun way to carry out this activity.

- Practice "old" skills (15 minutes).
  - o Explain or remind participants of the rationale for practicing skills that have already been learned—need practice to refine them, build efficiency, explore various ways to perform them, and generalize them to different settings.
  - o Generally using a random order for practicing the old skills is ideal, but it is acceptable to begin with less stressful ones before proceeding to more difficult ones.
  - o Practice old skills in a variety of settings and using a variety of methods.
  - o For the old-skill practice, the trainer provides structure and safety, keeping feedback to a minimum.
  - o Individualize the difficulty level to the extent possible.
- Practice "new" skills (ones that have yet to be acquired or perfected) (15 minutes).
  - o This section of the session may carry over from one session to the next.
  - o The trainer should focus on a single skill or series of a few related skills.
  - o When introducing a new skill, the trainer should explain the rationale for the skill, demonstrate how it is done, then ask each participant to attempt the skill (either all at the same time or sequentially, depending upon the skill and the setting).
  - o Trainer role: Provide structure, safety, instructions, demonstration, and feedback.
- Warm-down activity (10 minutes) (optional if pressed for time).
  - o An activity designed to reduce any tension or frustration from working on the new skills.
  - o As for the warm-up activity, the activity should include some skills that have already been learned.
  - o A game can be a fun way to carry out this activity.
  - o Moving outside the regular training area (e.g., outside) can be useful.
- Closing (5 minutes).
  - o Form a circle ("huddle").
  - o Answer any questions that the learners may have.
  - o Summarize the key points about the "new" skills covered earlier in the session.
  - o If it is not the final session:
    - – Have each participant review his/her training goals, revising them if appropriate.
    - – Remind participants and training personnel of the date, time, and location for the next session.

- Strongly encourage participants to practice their skills (with a spotter if needed) before the next session.
o If it is the final session:
  - Review any arrangements for obtaining a post-training assessment of wheelchair skills (e.g., using the WST-Q or WST), preferably a minimum of 3 days after the final training session.
  - Congratulate participants on their participation and achievements.
  - Have participants complete an evaluation form on the training sessions.
  - Distribute report cards and certificates.
  - Award prizes—preferably some small trinket for each participant with a humorous reason (e.g., "For the best spotter scare," "For the best uphill slalom," "For the fastest downhill sprint," "For the wobbliest wheelie without falling").
  - Thank training personnel for their efforts.
o Retrieve materials (name tags, clip-boards, and pens or pencils) from the participants.
o Complete any final documentation of the session.

Overview Schedule of "Old" and "New" Skill Coverage

Individual Skills	Session #					
	1	2	3	4	5	6
1. Rolls forward short distance	New	Old				
2. Rolls backward short distance	New	Old				
3. Turns in place	New	Old				
4. Turns while moving forward	New	Old				
5. Turns while moving backward	New	Old				
6. Maneuvers sideways	New	Old				
7. Reaches high object		New	Old			
8. Picks object from floor		New	Old			
9. Relieves weight from buttocks		New	Old			
10. Operates body positioning options		New	Old			
11. Level transfer		New	Old			
12. Folds and unfolds wheelchair		New	Old			
13. Gets through hinged door		New	Old			
14. Rolls longer distance		New	Old			
15. Avoids moving obstacles		New	Old			
16. Ascends slight incline			New	Old		
17. Descends slight incline			New	Old		
18. Ascends steep incline			New	Old		
19. Descends steep incline			New	Old		
20. Rolls across side-slope			New	Old		
21. Rolls on soft surface			New	Old		
22. Gets over threshold				New	Old	
23. Gets over gap				New	Old	
24. Ascends low curb				New	Old	
25. Descends low curb				New	Old	
26. Ascends high curb				New	Old	
27. Descends high curb				New	Old	
28. Performs stationary wheelie					New	Old
29. Turns in place in wheelie position					New	Old
30. Descends high curb in wheelie position					New	Old
31. Descends steep incline in wheelie position					New	Old
32. Gets from ground into wheelchair						New
33. Ascends stairs						New
34. Descends stairs						New

A2.2 Large groups

Groups of up to 20 learners can be managed by 1–2 trainers for workshops lasting 1–8 hours on a single day. Typically for a group of 20, the whole group will work together on the same skills. For some skills (e.g., "rolls forward short distance"), a "conga line" can be used (each learner following the other by a few meters). For other skills (e.g., "maneuvers sideways"), a "chorus line" can be used (all learners performing the same skill at the same time).

Groups of more than 20 learners can be divided into smaller groups and managed on different days. Alternatively, stations can be used. For instance, a group of 24 therapy students undergoing a 90-minute workshop on wheelchair skills in a 2-hour period of time can be divided into 6 groups of 4 students with each group starting at a different station ("shotgun" start). At each station, one trainer deals with a different set of skills for 15 minutes, then the students move on to the next station.

If there is more flexibility with respect to the starting and finishing time (e.g., a 4-hour time period even though each student will only receive 90 minutes of training), it is preferable to use a "staggered" start, with each group beginning at Station 1 and progressing every 15 minutes in order to Station 6. This has the advantage over the shotgun start of allowing each student to learn about the skills in the preferred sequence.

When working with large groups of able-bodied learners (e.g., students in the health professions), the number of available wheelchairs can be a limiting factor. For a group of 20 able-bodied learners, 10 wheelchairs of different sizes are ideal, with 10 students in the wheelchairs and the other 10 acting as spotters at any time. If there are more students or fewer wheelchairs, then the students can be divided into group sizes that are a multiple of the number of wheelchairs. For instance, for 30 learners and 10 wheelchairs, there would be 3 groups of 10 students. For each skill, Group 1 students would begin in the wheelchairs, Group 2 students would be the spotters, and Group 3 students would be observers. After each student in Group 1 has attempted the skill, the Group 2 students get into the wheelchairs and the Group 3 students act as spotters. This rotation continues until every student has attempted every skill.

Appendix 3: WST report forms

Wheelchair Skills Test (WST) Version 4.3 Form

Manual Wheelchairs Operated by Their Users

Name of wheelchair user: _____

Tester: _____ Date: _____

#	Individual Skill	Capacity (0–2)	Comments
1	Rolls forward short distance		
2	Rolls backward short distance		
3	Turns in place		
4	Turns while moving forward		
5	Turns while moving backward		
6	Maneuvers sideways		
7	Reaches high object		
8	Picks object from floor		
9	Relieves weight from buttocks		
10	Operates body positioning options		
11	Level transfer		
12	Folds and unfolds wheelchair		
13	Gets through hinged door		
14	Rolls longer distance		
15	Avoids moving obstacles		
16	Ascends slight incline		
17	Descends slight incline		
18	Ascends steep incline		
19	Descends steep incline		
20	Rolls across side-slope		
21	Rolls on soft surface		
22	Gets over threshold		
23	Gets over gap		
24	Ascends low curb		
25	Descends low curb		
26	Ascends high curb		
27	Descends high curb		
28	Performs stationary wheelie		
29	Turns in place in wheelie position		
30	Descends high curb in wheelie position		
31	Descends steep incline in wheelie position		
32	Gets from ground into wheelchair		
33	Ascends stairs		
34	Descends stairs		
	Total score:	%	

General comments:

Training goals described by the wheelchair user:

Name and address of any person(s) to whom the test subject would like a copy of the report to be sent.

Details about the WST can be found in the WSP Manual at http://www.wheelchairskillsprogram.ca/eng/manual.php/.

Wheelchair Skills Test (WST) Version 4.3 Form

Manual Wheelchairs Operated by Caregivers

Name of caregiver: _____

Tester: _____ Date: _____

#	Individual Skill	Capacity (0–2)	Comments
1	Rolls forward short distance		
2	Rolls backward short distance		
3	Turns in place		
4	Turns while moving forward		
5	Turns while moving backward		
6	Maneuvers sideways		
7	Picks object from floor		
8	Relieves weight from buttocks		
9	Operates body positioning options		
10	Level transfer		
11	Folds and unfolds wheelchair		
12	Gets through hinged door		
13	Rolls longer distance		
14	Avoids moving obstacles		
15	Ascends slight incline		
16	Descends slight incline		
17	Ascends steep incline		
18	Descends steep incline		
19	Rolls across side-slope		
20	Rolls on soft surface		
21	Gets over threshold		
22	Gets over gap		
23	Ascends low curb		
24	Descends low curb		
25	Ascends high curb		
26	Descends high curb		
27	Performs stationary wheelie		
28	Turns in place in wheelie position		
29	Descends high curb in wheelie position		
30	Descends steep incline in wheelie position		
31	Gets from ground into wheelchair		
32	Ascends stairs		
33	Descends stairs		
	Total score:	%	

General comments:

Training goals described by the caregiver:

Name and address of any person(s) to whom the test subject would like a copy of the report to be sent.

Details about the WST can be found in the WSP Manual at http://www.wheel-chairskillsprogram.ca/eng/manual.php/.

Wheelchair Skills Test (WST) Version 4.3 Form

Powered Wheelchairs Operated by Their Users

Name of wheelchair user: _____

Tester: _____ Date: _____

#	Individual Skill	Capacity (0–2)	Comments
1	Moves controller away and back		
2	Turns power on and off		
3	Selects drive modes and speeds		
4	Disengages and engages motors		
5	Operates battery charger		
6	Rolls forward short distance		
7	Rolls backward short distance		
8	Turns in place		
9	Turns while moving forward		
10	Turns while moving backward		
11	Maneuvers sideways		
12	Reaches high object		
13	Picks object from floor		
14	Relieves weight from buttocks		
15	Operates body positioning options		
16	Level transfer		
17	Gets through hinged door		
18	Rolls longer distance		
19	Avoids moving obstacles		
20	Ascends slight incline		
21	Descends slight incline		
22	Ascends steep incline		
23	Descends steep incline		
24	Rolls across side-slope		
25	Rolls on soft surface		
26	Gets over threshold		
27	Gets over gap		
28	Ascends low curb		
29	Descends low curb		
30	Gets from ground into wheelchair		
	Total score:	%	

General comments:

Training goals described by the wheelchair user:

Name and address of any person(s) to whom the test subject would like a copy of the report to be sent.

Details about the WST can be found in the WSP Manual at http://www.wheel-chairskillsprogram.ca/eng/manual.php/.

Wheelchair Skills Test (WST) Version 4.3 Form

Powered Wheelchairs Operated by Caregivers

Name of caregiver: _____

Tester: _____ Date: _____

#	Individual Skill	Capacity (0–2)	Comments
1	Moves controller away and back		
2	Turns power on and off		
3	Selects drive modes and speeds		
4	Disengages and engages motors		
5	Operates battery charger		
6	Rolls forward short distance		
7	Rolls backward short distance		
8	Turns in place		
9	Turns while moving forward		
10	Turns while moving backward		
11	Maneuvers sideways		
12	Reaches high object		
13	Picks object from floor		
14	Operates body positioning options		
15	Level transfer		
16	Gets through hinged door		
17	Rolls longer distance		
18	Avoids moving obstacles		
19	Ascends slight incline		
20	Descends slight incline		
21	Ascends steep incline		
22	Descends steep incline		
23	Rolls across side-slope		
24	Rolls on soft surface		
25	Gets over threshold		
26	Gets over gap		
27	Ascends low curb		
28	Descends low curb		
29	Gets from ground into wheelchair		
	Total score:	%	

General comments:

Training goals described by the caregiver:

Name and address of any person(s) to whom the test subject would like a copy of the report to be sent.

Details about the WST can be found in the WSP Manual at http://www.wheelchairskillsprogram.ca/eng/manual.php/.

Wheelchair Skills Test (WST) Version 4.3 Form

Scooters Operated by Their Users

Name of scooter user: _____

Tester: _____ Date: _____

#	Individual Skill	Capacity (0–2)	Comments
1	Moves controller away and back		
2	Turns power on and off		
3	Selects drive modes and speeds		
4	Disengages and engages motors		
5	Operates battery charger		
6	Rolls forward short distance		
7	Rolls backward short distance		
8	Turns in place		
9	Turns while moving forward		
10	Turns while moving backward		
11	Maneuvers sideways		
12	Picks object from floor		
13	Relieves weight from buttocks		
14	Operates body positioning options		
15	Level transfer		
16	Gets through hinged door		
17	Rolls longer distance		
18	Avoids moving obstacles		
19	Ascends slight incline		
20	Descends slight incline		
21	Ascends steep incline		
22	Descends steep incline		
23	Rolls across side-slope		
24	Rolls on soft surface		
25	Gets over threshold		
26	Gets over gap		
27	Ascends low curb		
28	Descends low curb		
29	Gets from ground into scooter		
	Total score:	%	

General comments:

Training goals described by the scooter user:

Name and address of any person(s) to whom the test subject would like a copy
of the report to be sent.

Details about the WST can be found in the WSP Manual at http://www.wheel-
chairskillsprogram.ca/eng/manual.php/.

Appendix 4: WST-Q scripts

Wheelchair Skills Test Questionnaire (WST-Q), Version 4.3 for Manual Wheelchairs Operated by Their Users

Question	Answer
Name of the wheelchair user:	
Date questionnaire completed (month, day, year):	
Did you complete the questionnaire yourself?	☐ Yes ☐ No
If you had help, what is the name of the person who helped you?	
If you had help, what is the relationship between you and the person who helped you?	☐ Family member ☐ Friend ☐ Caregiver ☐ Other person

Introduction to the questionnaire

- Copies of this questionnaire can be downloaded from www.wheelchair-skillsprogram.ca/eng/wstq.php.
- More details about the questionnaire can be found earlier in this book or in the WSP Manual.
- In this questionnaire, you will be asked questions about different skills that you might do in your wheelchair. These skills range from ones that are more basic at the beginning to those that are more advanced at the end.
- There are no "right" or "wrong" answers. The purpose of the questionnaire is simply to help us understand how you use the wheelchair.
- It will probably take about 10 minutes to complete the questionnaire, but please take as much time as you need.
- If you have more than one wheelchair, the questions are about the wheelchair that you use most often.
- If you have any comments, you will be able to record them at the end of the questionnaire.
- For each specific skill, you will be asked up to four questions. The questions and the possible answers are shown below.
- For each skill, you should answer the following question:

Question: "Can you do it?"	
Possible answers	**What this means**
Yes	I can safely do the skill, by myself, without any difficulty.
Yes with difficulty	Yes, but not as well as I would like.
No	I have never done the skill or I do not feel that I could do it right now.
Not possible with this wheelchair	The wheelchair does not have the parts to allow this skill. (This option is only presented for skills where such a score is a possibility.)

- If one of the purposes of this questionnaire is to assess how confident you are in performing the skill, you should also answer the following question for each skill:

Confidence Question: "How confident are you?"	
Possible answers	What this means
Fully	As of now, I am fully confident that I can do this skill safely and consistently.
Somewhat	As of now, I am somewhat confident that I can do this skill safely and consistently.
Not at all	As of now, I am not at all confident that I can do this skill safely and consistently.

- If one of the purposes of this questionnaire is to assess how often you do the skill, you should also answer the following question for each skill:

Question: "How often do you do it?"	
Possible answers	What this means
Always	Whenever I need or want to do so.
Sometimes	Sometimes when I need or want to, sometimes not.
Never	Never or less often than once a year.

- If one of the purposes of this questionnaire is to identify goals for training, you should also answer the following question about each skill:

Question: "Is this a training goal?"	
Possible answers	What this means
Yes	I am interested in receiving training for this skill.
No	I am not interested in receiving training for this skill.

- If you have training goals that you can think of now, please record them in the space available below. You will have a chance to identify other goals later.

- Please read the questions about specific skills that begin on the next page. For each skill, record the answers in the spaces provided.

Questions on Specific Skills

		Questions (Pick only one answer for each question)			
#	Skill description	Can you do it?	How confident are you?	How often do you do it?	Is this a training goal?
1	Moving the wheelchair straight forward for a short distance, for example, along a short hallway.	☐ Yes ☐ Yes with difficulty ☐ No	☐ Fully ☐ Somewhat ☐ Not at all	☐ Always ☐ Sometimes ☐ Never	☐ Yes ☐ No
2	Moving the wheelchair straight backward for a short distance, for example, to back away from a table.	☐ Yes ☐ Yes with difficulty ☐ No	☐ Fully ☐ Somewhat ☐ Not at all	☐ Always ☐ Sometimes ☐ Never	☐ Yes ☐ No
3	Turning the wheelchair around in a small space so that it is facing in the opposite direction.	☐ Yes ☐ Yes with difficulty ☐ No	☐ Fully ☐ Somewhat ☐ Not at all	☐ Always ☐ Sometimes ☐ Never	☐ Yes ☐ No
4	Turning the wheelchair around a corner while moving forwards.	☐ Yes ☐ Yes with difficulty ☐ No	☐ Fully ☐ Somewhat ☐ Not at all	☐ Always ☐ Sometimes ☐ Never	☐ Yes ☐ No
5	Turning the wheelchair around a corner while moving backward.	☐ Yes ☐ Yes with difficulty ☐ No	☐ Fully ☐ Somewhat ☐ Not at all	☐ Always ☐ Sometimes ☐ Never	☐ Yes ☐ No
6	Moving the wheelchair sideways in a small space, for example, to get the side of the wheelchair next to a kitchen counter, and then back to where you started.	☐ Yes ☐ Yes with difficulty ☐ No	☐ Fully ☐ Somewhat ☐ Not at all	☐ Always ☐ Sometimes ☐ Never	☐ Yes ☐ No
7	Moving the wheelchair to reach up for something overhead, for example, a high elevator button.	☐ Yes ☐ Yes with difficulty ☐ No	☐ Fully ☐ Somewhat ☐ Not at all	☐ Always ☐ Sometimes ☐ Never	☐ Yes ☐ No
8	Moving the wheelchair to pick up a small object, for example, a paperback book, from the floor in front of you.	☐ Yes ☐ Yes with difficulty ☐ No	☐ Fully ☐ Somewhat ☐ Not at all	☐ Always ☐ Sometimes ☐ Never	☐ Yes ☐ No
9	Operating all of the positioning options of the wheelchair (e.g., tilting the seat, reclining the seat, elevating the leg-rests).	☐ Yes ☐ Yes with difficulty ☐ No ☐ Not possible with this wheelchair	☐ Fully ☐ Somewhat ☐ Not at all	☐ Always ☐ Sometimes ☐ Never	☐ Yes ☐ No

Continued

#	Skill description	Can you do it?	How confident are you?	How often do you do it?	Is this a training goal?
		Questions (Pick only one answer for each question)			
10	Removing the weight from your buttocks, either one at a time or both together.	☐ Yes ☐ Yes with difficulty ☐ No	☐ Fully ☐ Somewhat ☐ Not at all	☐ Always ☐ Sometimes ☐ Never	☐ Yes ☐ No
11	Transferring from the wheelchair to a bench that is about the same height as the wheelchair and then getting back into the wheelchair.	☐ Yes ☐ Yes with difficulty ☐ No	☐ Fully ☐ Somewhat ☐ Not at all	☐ Always ☐ Sometimes ☐ Never	☐ Yes ☐ No
12	Folding the wheelchair or taking it apart without tools, for example, to store it out of the way, and then opening or reassembling it again.	☐ Yes ☐ Yes with difficulty ☐ No ☐ Not possible with this wheelchair	☐ Fully ☐ Somewhat ☐ Not at all	☐ Always ☐ Sometimes ☐ Never	☐ Yes ☐ No
13	Opening a hinged door, moving the wheelchair through it, and closing it behind you, then coming back the other way.	☐ Yes ☐ Yes with difficulty ☐ No	☐ Fully ☐ Somewhat ☐ Not at all	☐ Always ☐ Sometimes ☐ Never	☐ Yes ☐ No
14	Moving the wheelchair over a longer distance, for example, on a smooth surface about the length of a sport field.	☐ Yes ☐ Yes with difficulty ☐ No	☐ Fully ☐ Somewhat ☐ Not at all	☐ Always ☐ Sometimes ☐ Never	☐ Yes ☐ No
15	While moving the wheelchair, avoiding moving people who do not notice you.	☐ Yes ☐ Yes with difficulty ☐ No	☐ Fully ☐ Somewhat ☐ Not at all	☐ Always ☐ Sometimes ☐ Never	☐ Yes ☐ No
16	Moving the wheelchair up a slight incline, for example, a standard ramp (12 times longer than it is high).	☐ Yes ☐ Yes with difficulty ☐ No	☐ Fully ☐ Somewhat ☐ Not at all	☐ Always ☐ Sometimes ☐ Never	☐ Yes ☐ No
17	Moving the wheelchair down a slight incline.	☐ Yes ☐ Yes with difficulty ☐ No	☐ Fully ☐ Somewhat ☐ Not at all	☐ Always ☐ Sometimes ☐ Never	☐ Yes ☐ No
18	Moving the wheelchair up a steep incline (about twice as steep as a standard ramp).	☐ Yes ☐ Yes with difficulty ☐ No	☐ Fully ☐ Somewhat ☐ Not at all	☐ Always ☐ Sometimes ☐ Never	☐ Yes ☐ No

Continued

#	Skill description	Questions (Pick only one answer for each question)			
		Can you do it?	How confident are you?	How often do you do it?	Is this a training goal?
19	Moving the wheelchair down a steep incline.	☐ Yes ☐ Yes with difficulty ☐ No	☐ Fully ☐ Somewhat ☐ Not at all	☐ Always ☐ Sometimes ☐ Never	☐ Yes ☐ No
20	Moving the wheelchair across a slight side-slope, for example, when crossing a driveway.	☐ Yes ☐ Yes with difficulty ☐ No	☐ Fully ☐ Somewhat ☐ Not at all	☐ Always ☐ Sometimes ☐ Never	☐ Yes ☐ No
21	Moving the wheelchair a short distance across a soft surface, for example, gravel.	☐ Yes ☐ Yes with difficulty ☐ No	☐ Fully ☐ Somewhat ☐ Not at all	☐ Always ☐ Sometimes ☐ Never	☐ Yes ☐ No
22	Getting the wheelchair over an obstacle that sticks up above the surface, for example, a door threshold.	☐ Yes ☐ Yes with difficulty ☐ No	☐ Fully ☐ Somewhat ☐ Not at all	☐ Always ☐ Sometimes ☐ Never	☐ Yes ☐ No
23	Getting the wheelchair over a gap, for example, a rut in the road that is too big to simply roll over.	☐ Yes ☐ Yes with difficulty ☐ No	☐ Fully ☐ Somewhat ☐ Not at all	☐ Always ☐ Sometimes ☐ Never	☐ Yes ☐ No
24	Getting the wheelchair up a low curb, for example, when entering a building.	☐ Yes ☐ Yes with difficulty ☐ No	☐ Fully ☐ Somewhat ☐ Not at all	☐ Always ☐ Sometimes ☐ Never	☐ Yes ☐ No
25	Getting the wheelchair down from a low curb.	☐ Yes ☐ Yes with difficulty ☐ No	☐ Fully ☐ Somewhat ☐ Not at all	☐ Always ☐ Sometimes ☐ Never	☐ Yes ☐ No
26	Getting the wheelchair up a high curb, for example, at a street corner without a ramp.	☐ Yes ☐ Yes with difficulty ☐ No	☐ Fully ☐ Somewhat ☐ Not at all	☐ Always ☐ Sometimes ☐ Never	☐ Yes ☐ No
27	Getting the wheelchair down from a high curb.	☐ Yes ☐ Yes with difficulty ☐ No	☐ Fully ☐ Somewhat ☐ Not at all	☐ Always ☐ Sometimes ☐ Never	☐ Yes ☐ No
28	Doing a wheelie, balancing the wheelchair on its rear wheels, for 30 seconds.	☐ Yes ☐ Yes with difficulty ☐ No	☐ Fully ☐ Somewhat ☐ Not at all	☐ Always ☐ Sometimes ☐ Never	☐ Yes ☐ No
29	Staying in a wheelie, turning the wheelchair around in a small space so that it is facing in the opposite direction.	☐ Yes ☐ Yes with difficulty ☐ No	☐ Fully ☐ Somewhat ☐ Not at all	☐ Always ☐ Sometimes ☐ Never	☐ Yes ☐ No

Continued

#	Skill description	Questions (Pick only one answer for each question)			
		Can you do it?	How confident are you?	How often do you do it?	Is this a training goal?
30	Staying in a wheelie, moving forward down a high curb.	☐ Yes ☐ Yes with difficulty ☐ No	☐ Fully ☐ Somewhat ☐ Not at all	☐ Always ☐ Sometimes ☐ Never	☐ Yes ☐ No
31	Staying in a wheelie, moving forward down a steep ramp.	☐ Yes ☐ Yes with difficulty ☐ No	☐ Fully ☐ Somewhat ☐ Not at all	☐ Always ☐ Sometimes ☐ Never	☐ Yes ☐ No
32	Getting up from the ground into the wheelchair, for example, after a fall.	☐ Yes ☐ Yes with difficulty ☐ No	☐ Fully ☐ Somewhat ☐ Not at all	☐ Always ☐ Sometimes ☐ Never	☐ Yes ☐ No
33	Getting yourself and the wheelchair up a short flight of stairs that has a rail	☐ Yes ☐ Yes with difficulty ☐ No	☐ Fully ☐ Somewhat ☐ Not at all	☐ Always ☐ Sometimes ☐ Never	☐ Yes ☐ No
34	Getting yourself and the wheelchair down a short flight of stairs that has a rail	☐ Yes ☐ Yes with difficulty ☐ No	☐ Fully ☐ Somewhat ☐ Not at all	☐ Always ☐ Sometimes ☐ Never	☐ Yes ☐ No

If you have any general comments about the questions that you have answered above, please record them in the space available below.

```



```

If you have any training goals that you have not already mentioned, please record them in the space available below.

```



```

A short report form will be created from the answers that you have given. If you would like a copy of the report form for yourself or someone else, please record in the space available below the name and address of the person to whom the report should be sent.

```



```

This is the end of the questionnaire. Thank you for completing it.

Wheelchair Skills Test Questionnaire (WST-Q), Version 4.3 for Manual Wheelchairs Operated by Caregivers

Question	Answer
Name of the caregiver:	
Date questionnaire completed (month, day, year):	
Did you complete the questionnaire yourself?	☐ Yes ☐ No
If you had help, what is the name of the person who helped you?	
If you had help, what is the relationship between you and the person who helped you?	☐ Family member ☐ Friend ☐ Other person

Introduction to the questionnaire
- Copies of this questionnaire can be downloaded from www.wheelchair-skillsprogram.ca/eng/wstq.php.
- More details about the questionnaire can be found earlier in this book or in the WSP Manual.
- In this questionnaire, you will be asked questions about different skills that you might do in your role as the caregiver of a wheelchair user. These skills range from ones that are more basic at the beginning to those that are more advanced at the end.
- There are no "right" or "wrong" answers. The purpose of the questionnaire is simply to help us understand how you perform caregiver wheelchair skills.
- It will probably take about 10 minutes to complete the questionnaire, but please take as much time as you need.
- If the wheelchair user who you care for has more than one wheelchair, the questions are about the wheelchair that he/she uses most often.
- If you have any comments, you will be able to record them at the end of the questionnaire.
- For each specific skill, you will be asked up to four questions. The questions and the possible answers are shown below.
- For each skill, you should answer the following question:

Question: "Can you do it?"	
Possible answers	**What this means**
Yes	I can safely do the skill, by myself, without any difficulty.
Yes with difficulty	Yes, but not as well as I would like.
No	I have never done the skill or I do not feel that I could do it right now.
Not possible with this wheelchair	The wheelchair does not have the parts to allow this skill. (This option is only presented for skills where such a score is a possibility.)

- If one of the purposes of this questionnaire is to assess how confident you are in performing the skill, you should also answer the following question for each skill:

Confidence Question: "How confident are you?"	
Possible Answers	**What this means**
Fully	As of now, I am fully confident that I can do this skill safely and consistently.
Somewhat	As of now, I am somewhat confident that I can do this skill safely and consistently.
Not at all	As of now, I am not at all confident that I can do this skill safely and consistently.

- If one of the purposes of this questionnaire is to assess how often you do the skill, you should also answer the following question for each skill:

Question: "How often do you do it?"	
Possible answers	**What this means**
Always	Whenever I need or want to do so.
Sometimes	Sometimes when I need or want to, sometimes not.
Never	Never or less often than once a year.

- If one of the purposes of this questionnaire is to identify goals for training, you should also answer the following question about each skill:

Question: "Is this a training goal?"	
Possible answers	**What this means**
Yes	I am interested in receiving training for this skill.
No	I am not interested in receiving training for this skill.

- If you have training goals that you can think of now, please record them in the space available below. You will have a chance to identify other goals later.

- Please read the questions about specific skills that begin on the next page. For each skill, record the answers in the spaces provided.

Questions on Specific Skills

#	Skill description	Can you do it?	How confident are you?	How often do you do it?	Is this a training goal?
			Questions (Pick only one answer for each question)		
1	Moving the wheelchair straight forward for a short distance, for example, along a short hallway.	☐ Yes ☐ Yes with difficulty ☐ No	☐ Fully ☐ Somewhat ☐ Not at all	☐ Always ☐ Sometimes ☐ Never	☐ Yes ☐ No
2	Moving the wheelchair straight backward for a short distance, for example, to back away from a table.	☐ Yes ☐ Yes with difficulty ☐ No	☐ Fully ☐ Somewhat ☐ Not at all	☐ Always ☐ Sometimes ☐ Never	☐ Yes ☐ No
3	Turning the wheelchair around in a small space so that it is facing in the opposite direction.	☐ Yes ☐ Yes with difficulty ☐ No	☐ Fully ☐ Somewhat ☐ Not at all	☐ Always ☐ Sometimes ☐ Never	☐ Yes ☐ No
4	Turning the wheelchair around a corner while moving forward.	☐ Yes ☐ Yes with difficulty ☐ No	☐ Fully ☐ Somewhat ☐ Not at all	☐ Always ☐ Sometimes ☐ Never	☐ Yes ☐ No
5	Turning the wheelchair around a corner while moving backward.	☐ Yes ☐ Yes with difficulty ☐ No	☐ Fully ☐ Somewhat ☐ Not at all	☐ Always ☐ Sometimes ☐ Never	☐ Yes ☐ No
6	Moving the wheelchair sideways in a small space, for example, to get the side of the wheelchair next to a kitchen counter, and then back to where you started.	☐ Yes ☐ Yes with difficulty ☐ No	☐ Fully ☐ Somewhat ☐ Not at all	☐ Always ☐ Sometimes ☐ Never	☐ Yes ☐ No
7	Moving the wheelchair to pick up a small object, for example, a paperback book, from the floor in front of you.	☐ Yes ☐ Yes with difficulty ☐ No	☐ Fully ☐ Somewhat ☐ Not at all	☐ Always ☐ Sometimes ☐ Never	☐ Yes ☐ No
8	Removing the weight from the wheelchair user's buttocks, either one at a time or both together.	☐ Yes ☐ Yes with difficulty ☐ No	☐ Fully ☐ Somewhat ☐ Not at all	☐ Always ☐ Sometimes ☐ Never	☐ Yes ☐ No

Continued

#	Skill description	Questions (Pick only one answer for each question)			
		Can you do it?	How confident are you?	How often do you do it?	Is this a training goal?
9	Operating all of the positioning options of the wheelchair (e.g., tilting the seat, reclining the seat, elevating the leg-rests).	☐ Yes ☐ Yes with difficulty ☐ No ☐ Not possible with this wheelchair	☐ Fully ☐ Somewhat ☐ Not at all	☐ Always ☐ Sometimes ☐ Never	☐ Yes ☐ No
10	Transferring the wheelchair user from the wheelchair to a bench that is about the same height as the wheelchair and then getting him/her back into the wheelchair.	☐ Yes ☐ Yes with difficulty ☐ No	☐ Fully ☐ Somewhat ☐ Not at all	☐ Always ☐ Sometimes ☐ Never	☐ Yes ☐ No
11	Folding the wheelchair or taking it apart without tools, for example, to store it out of the way, and then opening or reassembling it again.	☐ Yes ☐ Yes with difficulty ☐ No ☐ Not possible with this wheelchair	☐ Fully ☐ Somewhat ☐ Not at all	☐ Always ☐ Sometimes ☐ Never	☐ Yes ☐ No
12	Opening a hinged door, moving the wheelchair through it and closing it behind you, then coming back the other way.	☐ Yes ☐ Yes with difficulty ☐ No	☐ Fully ☐ Somewhat ☐ Not at all	☐ Always ☐ Sometimes ☐ Never	☐ Yes ☐ No
13	Moving the wheelchair over a longer distance, for example, on a smooth surface about the length of a sport field.	☐ Yes ☐ Yes with difficulty ☐ No	☐ Fully ☐ Somewhat ☐ Not at all	☐ Always ☐ Sometimes ☐ Never	☐ Yes ☐ No
14	While moving the wheelchair, avoiding moving people who do not notice you.	☐ Yes ☐ Yes with difficulty ☐ No	☐ Fully ☐ Somewhat ☐ Not at all	☐ Always ☐ Sometimes ☐ Never	☐ Yes ☐ No
15	Moving the wheelchair up a slight incline, for example, a standard ramp (12 times longer than it is high).	☐ Yes ☐ Yes with difficulty ☐ No	☐ Fully ☐ Somewhat ☐ Not at all	☐ Always ☐ Sometimes ☐ Never	☐ Yes ☐ No

Continued

#	Skill description	Can you do it?	How confident are you?	How often do you do it?	Is this a training goal?
				Questions (Pick only one answer for each question)	
16	Moving the wheelchair down a slight incline.	☐ Yes ☐ Yes with difficulty ☐ No	☐ Fully ☐ Somewhat ☐ Not at all	☐ Always ☐ Sometimes ☐ Never	☐ Yes ☐ No
17	Moving the wheelchair up a steep incline (about twice as steep as a standard ramp).	☐ Yes ☐ Yes with difficulty ☐ No	☐ Fully ☐ Somewhat ☐ Not at all	☐ Always ☐ Sometimes ☐ Never	☐ Yes ☐ No
18	Moving the wheelchair down a steep incline.	☐ Yes ☐ Yes with difficulty ☐ No	☐ Fully ☐ Somewhat ☐ Not at all	☐ Always ☐ Sometimes ☐ Never	☐ Yes ☐ No
19	Moving the wheelchair across a slight side-slope, for example, when crossing a driveway.	☐ Yes ☐ Yes with difficulty ☐ No	☐ Fully ☐ Somewhat ☐ Not at all	☐ Always ☐ Sometimes ☐ Never	☐ Yes ☐ No
20	Moving the wheelchair a short distance across a soft surface, for example, gravel.	☐ Yes ☐ Yes with difficulty ☐ No	☐ Fully ☐ Somewhat ☐ Not at all	☐ Always ☐ Sometimes ☐ Never	☐ Yes ☐ No
21	Getting the wheelchair over an obstacle that sticks up above the surface, for example, a door threshold.	☐ Yes ☐ Yes with difficulty ☐ No	☐ Fully ☐ Somewhat ☐ Not at all	☐ Always ☐ Sometimes ☐ Never	☐ Yes ☐ No
22	Getting the wheelchair over a gap, for example, a rut in the road that is too big to simply roll over.	☐ Yes ☐ Yes with difficulty ☐ No	☐ Fully ☐ Somewhat ☐ Not at all	☐ Always ☐ Sometimes ☐ Never	☐ Yes ☐ No
23	Getting the wheelchair up a low curb, for example, when entering a building.	☐ Yes ☐ Yes with difficulty ☐ No	☐ Fully ☐ Somewhat ☐ Not at all	☐ Always ☐ Sometimes ☐ Never	☐ Yes ☐ No
24	Getting the wheelchair down from a low curb.	☐ Yes ☐ Yes with difficulty ☐ No	☐ Fully ☐ Somewhat ☐ Not at all	☐ Always ☐ Sometimes ☐ Never	☐ Yes ☐ No
25	Getting the wheelchair up a high curb, for example, at a street corner without a ramp.	☐ Yes ☐ Yes with difficulty ☐ No	☐ Fully ☐ Somewhat ☐ Not at all	☐ Always ☐ Sometimes ☐ Never	☐ Yes ☐ No

Continued

Appendix 4

#	Skill description	Can you do it?	How confident are you?	How often do you do it?	Is this a training goal?
26	Getting the wheelchair down from a high curb.	☐ Yes ☐ Yes with difficulty ☐ No	☐ Fully ☐ Somewhat ☐ Not at all	☐ Always ☐ Sometimes ☐ Never	☐ Yes ☐ No
27	Doing a wheelie, balancing the wheelchair on its rear wheels, for 30 seconds.	☐ Yes ☐ Yes with difficulty ☐ No	☐ Fully ☐ Somewhat ☐ Not at all	☐ Always ☐ Sometimes ☐ Never	☐ Yes ☐ No
28	Staying in a wheelie, turning the wheelchair around in a small space so that it is facing in the opposite direction.	☐ Yes ☐ Yes with difficulty ☐ No	☐ Fully ☐ Somewhat ☐ Not at all	☐ Always ☐ Sometimes ☐ Never	☐ Yes ☐ No
29	Staying in a wheelie, moving forward down a steep ramp.	☐ Yes ☐ Yes with difficulty ☐ No	☐ Fully ☐ Somewhat ☐ Not at all	☐ Always ☐ Sometimes ☐ Never	☐ Yes ☐ No
30	Staying in a wheelie, moving forward down a high curb.	☐ Yes ☐ Yes with difficulty ☐ No	☐ Fully ☐ Somewhat ☐ Not at all	☐ Always ☐ Sometimes ☐ Never	☐ Yes ☐ No
31	Getting the wheelchair user up from the ground into the wheelchair, for example, after a fall.	☐ Yes ☐ Yes with difficulty ☐ No	☐ Fully ☐ Somewhat ☐ Not at all	☐ Always ☐ Sometimes ☐ Never	☐ Yes ☐ No
32	Getting the wheelchair user and the wheelchair up a short flight of stairs that has a rail.	☐ Yes ☐ Yes with difficulty ☐ No	☐ Fully ☐ Somewhat ☐ Not at all	☐ Always ☐ Sometimes ☐ Never	☐ Yes ☐ No
33	Getting the wheelchair user and the wheelchair down a short flight of stairs that has a rail.	☐ Yes ☐ Yes with difficulty ☐ No	☐ Fully ☐ Somewhat ☐ Not at all	☐ Always ☐ Sometimes ☐ Never	☐ Yes ☐ No

Questions (Pick only one answer for each question)

374

If you have any general comments about the questions that you have answered above, please record them in the space available below.

```
┌─────────────────────────────────────────────────────────────┐
│                                                               │
│                                                               │
│                                                               │
│                                                               │
└─────────────────────────────────────────────────────────────┘
```

If you have any training goals that you have not already mentioned, please record them in the space available below.

```
┌─────────────────────────────────────────────────────────────┐
│                                                               │
│                                                               │
│                                                               │
│                                                               │
└─────────────────────────────────────────────────────────────┘
```

A short report form will be created from the answers that you have given. If you would like a copy of the report form for yourself or someone else, please record in the space available below the name and address of the person to whom the report should be sent.

```
┌─────────────────────────────────────────────────────────────┐
│                                                               │
│                                                               │
│                                                               │
│                                                               │
└─────────────────────────────────────────────────────────────┘
```

This is the end of the questionnaire. Thank you for completing it.

Wheelchair Skills Test Questionnaire (WST-Q), Version 4.3 for Powered Wheelchairs Operated by Their Users

Question	Answer
Name of the wheelchair user:	
Date questionnaire completed (month, day, year):	
Did you complete the questionnaire yourself?	☐ Yes ☐ No
If you had help, what is the name of the person who helped you?	
If you had help, what is the relationship between you and the person who helped you?	☐ Family member ☐ Friend ☐ Caregiver ☐ Other person

Introduction to the questionnaire
- Copies of this questionnaire can be downloaded from www.wheelchair-skillsprogram.ca/eng/wstq.php.
- More details about the questionnaire can be found earlier in this book or in the WSP Manual.
- In this questionnaire, you will be asked questions about different skills that you might do in your wheelchair. These skills range from ones that are more basic at the beginning to those that are more advanced at the end.
- There are no "right" or "wrong" answers. The purpose of the questionnaire is simply to help us understand how you use your wheelchair.
- It will probably take about 10 minutes to complete the questionnaire, but please take as much time as you need.
- If you have more than one wheelchair, the questions are about the wheelchair that you use most often.
- If you have any comments, you will be able to record them at the end of the questionnaire.
- For each specific skill, you will be asked up to four questions. The questions and the possible answers are shown below.
- For each skill, you should answer the following question:

Question: "Can you do it?"	
Possible answers	**What this means**
Yes	I can safely do the skill, by myself, without any difficulty.
Yes with difficulty	Yes, but not as well as I would like.
No	I have never done the skill or I do not feel that I could do it right now.
Not possible with this wheelchair	The wheelchair does not have the parts to allow this skill. (This option is only presented for skills where such a score is a possibility.)

- If one of the purposes of this questionnaire is to assess how confident you are in performing the skill, you should also answer the following question for each skill:

Confidence Question: "How confident are you?"	
Possible answers	What this means
Fully	As of now, I am fully confident that I can do this skill safely and consistently.
Somewhat	As of now, I am somewhat confident that I can do this skill safely and consistently.
Not at all	As of now, I am not at all confident that I can do this skill safely and consistently.

- If one of the purposes of this questionnaire is to assess how often you do the skill, you should also answer the following question for each skill:

Question: "How often do you do it?"	
Possible answers	What this means
Always	Whenever I need or want to do so.
Sometimes	Sometimes when I need or want to, sometimes not.
Never	Never or less often than once a year.

- If one of the purposes of this questionnaire is to identify goals for training, you should also answer the following question about each skill:

Question: "Is this a training goal?"	
Possible answers	What This Means
Yes	I am interested in receiving training for this skill.
No	I am not interested in receiving training for this skill.

- If you have training goals that you can think of now, please record them in the space available below. You will have a chance to identify other goals later.

- Please read the questions about specific skills that begin on the next page. For each skill, record the answers in the spaces provided.

Questions on Specific Skills

#	Skill description	Can you do it?	How confident are you?	How often do you do it?	Is this a training goal?
		Questions (Pick only one answer for each question)			
1	Moving the controller away and back again.	☐ Yes ☐ Yes with difficulty ☐ No ☐ Not possible with this wheelchair	☐ Fully ☐ Somewhat ☐ Not at all	☐ Always ☐ Sometimes ☐ Never	☐ Yes ☐ No
2	Turning the power for the wheelchair on and off.	☐ Yes ☐ Yes with difficulty ☐ No	☐ Fully ☐ Somewhat ☐ Not at all	☐ Always ☐ Sometimes ☐ Never	☐ Yes ☐ No
3	Changing the settings and speeds for the wheelchair.	☐ Yes ☐ Yes with difficulty ☐ No ☐ Not possible with this wheelchair	☐ Fully ☐ Somewhat ☐ Not at all	☐ Always ☐ Sometimes ☐ Never	☐ Yes ☐ No
4	Disengaging the motors of the wheelchair, so that it can be pushed without power, and then engaging the motors again.	☐ Yes ☐ Yes with difficulty ☐ No	☐ Fully ☐ Somewhat ☐ Not at all	☐ Always ☐ Sometimes ☐ Never	☐ Yes ☐ No
5	Charging the battery for the wheelchair.	☐ Yes ☐ Yes with difficulty ☐ No	☐ Fully ☐ Somewhat ☐ Not at all	☐ Always ☐ Sometimes ☐ Never	☐ Yes ☐ No
6	Moving the wheelchair straight forwards for a short distance, for example, along a short hallway.	☐ Yes ☐ Yes with difficulty ☐ No	☐ Fully ☐ Somewhat ☐ Not at all	☐ Always ☐ Sometimes ☐ Never	☐ Yes ☐ No
7	Moving the wheelchair straight backwards for a short distance, for example, to back away from a table.	☐ Yes ☐ Yes with difficulty ☐ No	☐ Fully ☐ Somewhat ☐ Not at all	☐ Always ☐ Sometimes ☐ Never	☐ Yes ☐ No
8	Turning the wheelchair around in a small space so that it is facing in the opposite direction.	☐ Yes ☐ Yes with difficulty ☐ No	☐ Fully ☐ Somewhat ☐ Not at all	☐ Always ☐ Sometimes ☐ Never	☐ Yes ☐ No
9	Turning the wheelchair around a corner while moving forwards.	☐ Yes ☐ Yes with difficulty ☐ No	☐ Fully ☐ Somewhat ☐ Not at all	☐ Always ☐ Sometimes ☐ Never	☐ Yes ☐ No

Continued

#	Skill description	Questions (Pick only one answer for each question)			
		Can you do it?	How confident are you?	How often do you do it?	Is this a training goal?
10	Turning the wheelchair around a corner while moving backwards.	☐ Yes ☐ Yes with difficulty ☐ No	☐ Fully ☐ Somewhat ☐ Not at all	☐ Always ☐ Sometimes ☐ Never	☐ Yes ☐ No
11	Moving the wheelchair sideways in a small space, for example, to get the side of the wheelchair next to a kitchen counter, and then back to where you started.	☐ Yes ☐ Yes with difficulty ☐ No	☐ Fully ☐ Somewhat ☐ Not at all	☐ Always ☐ Sometimes ☐ Never	☐ Yes ☐ No
12	Moving the wheelchair to reach up for something overhead, for example, a high elevator button.	☐ Yes ☐ Yes with difficulty ☐ No	☐ Fully ☐ Somewhat ☐ Not at all	☐ Always ☐ Sometimes ☐ Never	☐ Yes ☐ No
13	Moving the wheelchair to pick up a small object, for example, a paperback book, from the floor in front of you.	☐ Yes ☐ Yes with difficulty ☐ No	☐ Fully ☐ Somewhat ☐ Not at all	☐ Always ☐ Sometimes ☐ Never	☐ Yes ☐ No
14	Removing the weight from your buttocks, either one at a time or both together.	☐ Yes ☐ Yes with difficulty ☐ No	☐ Fully ☐ Somewhat ☐ Not at all	☐ Always ☐ Sometimes ☐ Never	☐ Yes ☐ No
15	Operating all of the positioning options of the wheelchair (e.g., tilting the seat, reclining the seat, elevating the leg-rests).	☐ Yes ☐ Yes with difficulty ☐ No ☐ Not possible with this wheelchair	☐ Fully ☐ Somewhat ☐ Not at all	☐ Always ☐ Sometimes ☐ Never	☐ Yes ☐ No
16	Transferring from the wheelchair to a bench that is about the same height as the wheelchair and then getting back into the wheelchair.	☐ Yes ☐ Yes with difficulty ☐ No	☐ Fully ☐ Somewhat ☐ Not at all	☐ Always ☐ Sometimes ☐ Never	☐ Yes ☐ No
17	Opening a hinged door, moving the wheelchair through it, and closing it behind you, then coming back the other way.	☐ Yes ☐ Yes with difficulty ☐ No	☐ Fully ☐ Somewhat ☐ Not at all	☐ Always ☐ Sometimes ☐ Never	☐ Yes ☐ No

Continued

#	Skill description	Questions (Pick only one answer for each question)			
		Can you do it?	How confident are you?	How often do you do it?	Is this a training goal?
18	Moving the wheelchair over a longer distance, for example, on a smooth surface about the length of a sport field.	☐ Yes ☐ Yes with difficulty ☐ No	☐ Fully ☐ Somewhat ☐ Not at all	☐ Always ☐ Sometimes ☐ Never	☐ Yes ☐ No
19	While moving the wheelchair, avoiding moving people who do not notice you.	☐ Yes ☐ Yes with difficulty ☐ No	☐ Fully ☐ Somewhat ☐ Not at all	☐ Always ☐ Sometimes ☐ Never	☐ Yes ☐ No
20	Moving the wheelchair up a slight incline, for example, a standard ramp (12 times longer than it is high).	☐ Yes ☐ Yes with difficulty ☐ No	☐ Fully ☐ Somewhat ☐ Not at all	☐ Always ☐ Sometimes ☐ Never	☐ Yes ☐ No
21	Moving the wheelchair down a slight incline.	☐ Yes ☐ Yes with difficulty ☐ No	☐ Fully ☐ Somewhat ☐ Not at all	☐ Always ☐ Sometimes ☐ Never	☐ Yes ☐ No
22	Moving the wheelchair up a steep incline (about twice as steep as a standard ramp).	☐ Yes ☐ Yes with difficulty ☐ No	☐ Fully ☐ Somewhat ☐ Not at all	☐ Always ☐ Sometimes ☐ Never	☐ Yes ☐ No
23	Moving the wheelchair down a steep incline.	☐ Yes ☐ Yes with difficulty ☐ No	☐ Fully ☐ Somewhat ☐ Not at all	☐ Always ☐ Sometimes ☐ Never	☐ Yes ☐ No
24	Moving the wheelchair across a slight side-slope, for example, when crossing a driveway.	☐ Yes ☐ Yes with difficulty ☐ No	☐ Fully ☐ Somewhat ☐ Not at all	☐ Always ☐ Sometimes ☐ Never	☐ Yes ☐ No
25	Moving the wheelchair a short distance across a soft surface, for example, gravel.	☐ Yes ☐ Yes with difficulty ☐ No	☐ Fully ☐ Somewhat ☐ Not at all	☐ Always ☐ Sometimes ☐ Never	☐ Yes ☐ No
26	Getting the wheelchair over an obstacle that sticks up above the surface, for example, a door threshold.	☐ Yes ☐ Yes with difficulty ☐ No	☐ Fully ☐ Somewhat ☐ Not at all	☐ Always ☐ Sometimes ☐ Never	☐ Yes ☐ No
27	Getting the wheelchair over a gap, for example, a rut in the road that is too big to simply roll over.	☐ Yes ☐ Yes with difficulty ☐ No	☐ Fully ☐ Somewhat ☐ Not at all	☐ Always ☐ Sometimes ☐ Never	☐ Yes ☐ No

Continued

#	Skill description	Questions (Pick only one answer for each question)			
		Can you do it?	How confident are you?	How often do you do it?	Is this a training goal?
28	Getting the wheelchair up a low curb, for example, when entering a building.	☐ Yes ☐ Yes with difficulty ☐ No	☐ Fully ☐ Somewhat ☐ Not at all	☐ Always ☐ Sometimes ☐ Never	☐ Yes ☐ No
29	Getting the wheelchair down from a low curb.	☐ Yes ☐ Yes with difficulty ☐ No	☐ Fully ☐ Somewhat ☐ Not at all	☐ Always ☐ Sometimes ☐ Never	☐ Yes ☐ No
30	Getting up from the ground into the wheelchair, for example, after a fall.	☐ Yes ☐ Yes with difficulty ☐ No	☐ Fully ☐ Somewhat ☐ Not at all	☐ Always ☐ Sometimes ☐ Never	☐ Yes ☐ No

If you have any general comments about the questions that you have answered above, please record them in the space available below.

```



```

If you have any training goals that you have not already mentioned, please record them in the space available below.

```



```

A short report form will be created from the answers that you have given. If you would like a copy of the report form for yourself or someone else, please record in the space available below the name and address of the person to whom the report should be sent.

```



```

This is the end of the questionnaire. Thank you for completing it.

Wheelchair Skills Test Questionnaire (WST-Q), Version 4.3 for Powered Wheelchairs Operated by Caregivers

Question	Answer
Name of the caregiver:	
Date questionnaire completed (month, day, year):	
Did you complete the questionnaire yourself?	☐ Yes ☐ No
If you had help, what is the name of the person who helped you?	
If you had help, what is the relationship between you and the person who helped you?	☐ Family member ☐ Friend ☐ Other person

Introduction to the questionnaire
- Copies of this questionnaire can be downloaded from www.wheelchair-skillsprogram.ca/eng/wstq.php.
- More details about the questionnaire can be found earlier in this book or in the WSP Manual.
- In this questionnaire, you will be asked questions about different skills that you might do in your role as the caregiver of a wheelchair user. These skills range from ones that are more basic at the beginning to those that are more advanced at the end.
- There are no "right" or "wrong" answers. The purpose of the questionnaire is simply to help us understand how you perform caregiver wheelchair skills.
- It will probably take about 10 minutes to complete the questionnaire, but please take as much time as you need.
- If the wheelchair user who you care for has more than one wheelchair, the questions are about the wheelchair that he/she uses most often.
- If you have any comments, you will be able to record them at the end of the questionnaire.
- For each specific skill, you will be asked up to four questions. The questions and the possible answers are shown below.
- For each skill, you should answer the following question:

Question: "Can you do it?"	
Possible answers	**What this means**
Yes	I can safely do the skill, by myself, without any difficulty.
Yes with difficulty	Yes, but not as well as I would like.
No	I have never done the skill or I do not feel that I could do it right now.
Not possible with this wheelchair	The wheelchair does not have the parts to allow this skill. (This option is only presented for skills where such a score is a possibility.)

- If one of the purposes of this questionnaire is to assess how confident you are in performing the skill, you should also answer the following question for each skill:

Confidence Question: "How confident are you?"	
Possible answers	**What this means**
Fully	As of now, I am fully confident that I can do this skill safely and consistently.
Somewhat	As of now, I am somewhat confident that I can do this skill safely and consistently.
Not at all	As of now, I am not at all confident that I can do this skill safely and consistently.

- If one of the purposes of this questionnaire is to assess how often you do the skill, you should also answer the following question for each skill:

Question: "How often do you do it?"	
Possible answers	**What this means**
Always	Whenever I need or want to do so.
Sometimes	Sometimes when I need or want to, sometimes not.
Never	Never or less often than once a year.

- If one of the purposes of this questionnaire is to identify goals for training, you should also answer the following question about each skill:

Question: "Is this a training goal?"	
Possible answers	**What this means**
Yes	I am interested in receiving training for this skill.
No	I am not interested in receiving training for this skill.

- If you have training goals that you can think of now, please record them in the space available below. You will have a chance to identify other goals later.

- Please read the questions about specific skills that begin on the next page. For each skill, record the answers in the spaces provided.

Questions on Specific Skills

#	Skill description	Questions (Pick only one answer for each question)			
		Can you do it?	How confident are you?	How often do you do it?	Is this a training goal?
1	Moving the controller away and back again.	□ Yes □ Yes with difficulty □ No □ Not possible with this wheelchair	□ Fully □ Somewhat □ Not at all	□ Always □ Sometimes □ Never	□ Yes □ No
2	Turning the power for the wheelchair on and off.	□ Yes □ Yes with difficulty □ No	□ Fully □ Somewhat □ Not at all	□ Always □ Sometimes □ Never	□ Yes □ No
3	Changing the settings and speeds for the wheelchair.	□ Yes □ Yes with difficulty □ No □ Not possible with this wheelchair	□ Fully □ Somewhat □ Not at all	□ Always □ Sometimes □ Never	□ Yes □ No
4	Disengaging the motors of the wheelchair, so that it can be pushed without power, and then engaging the motors again.	□ Yes □ Yes with difficulty □ No	□ Fully □ Somewhat □ Not at all	□ Always □ Sometimes □ Never	□ Yes □ No
5	Charging the battery for the wheelchair.	□ Yes □ Yes with difficulty □ No	□ Fully □ Somewhat □ Not at all	□ Always □ Sometimes □ Never	□ Yes □ No
6	Moving the wheelchair straight forward for a short distance, for example, along a short hallway.	□ Yes □ Yes with difficulty □ No	□ Fully □ Somewhat □ Not at all	□ Always □ Sometimes □ Never	□ Yes □ No
7	Moving the wheelchair straight backward for a short distance, for example, to back away from a table.	□ Yes □ Yes with difficulty □ No	□ Fully □ Somewhat □ Not at all	□ Always □ Sometimes □ Never	□ Yes □ No
8	Turning the wheelchair around in a small space so that it is facing in the opposite direction.	□ Yes □ Yes with difficulty □ No	□ Fully □ Somewhat □ Not at all	□ Always □ Sometimes □ Never	□ Yes □ No

Continued

#	Skill description	Can you do it?	How confident are you?	How often do you do it?	Is this a training goal?
		Questions (Pick only one answer for each question)			
9	Turning the wheelchair around a corner while moving forward.	☐ Yes ☐ Yes with difficulty ☐ No	☐ Fully ☐ Somewhat ☐ Not at all	☐ Always ☐ Sometimes ☐ Never	☐ Yes ☐ No
10	Turning the wheelchair around a corner while moving backward.	☐ Yes ☐ Yes with difficulty ☐ No	☐ Fully ☐ Somewhat ☐ Not at all	☐ Always ☐ Sometimes ☐ Never	☐ Yes ☐ No
11	Moving the wheelchair sideways in a small space, for example, to get the side of the wheelchair next to a kitchen counter, and then back to where you started.	☐ Yes ☐ Yes with difficulty ☐ No	☐ Fully ☐ Somewhat ☐ Not at all	☐ Always ☐ Sometimes ☐ Never	☐ Yes ☐ No
12	Moving the wheelchair to pick up a small object, for example, a paperback book, from the floor in front of you.	☐ Yes ☐ Yes with difficulty ☐ No	☐ Fully ☐ Somewhat ☐ Not at all	☐ Always ☐ Sometimes ☐ Never	☐ Yes ☐ No
13	Removing the weight from the wheelchair user's buttocks, either one at a time or both together.	☐ Yes ☐ Yes with difficulty ☐ No	☐ Fully ☐ Somewhat ☐ Not at all	☐ Always ☐ Sometimes ☐ Never	☐ Yes ☐ No
14	Operating all of the positioning options of the wheelchair (e.g., tilting the seat, reclining the seat, elevating the leg-rests).	☐ Yes ☐ Yes with difficulty ☐ No ☐ Not possible with this wheelchair	☐ Fully ☐ Somewhat ☐ Not at all	☐ Always ☐ Sometimes ☐ Never	☐ Yes ☐ No
15	Transferring the wheelchair user from the wheelchair to a bench that is about the same height as the wheelchair and then getting him/her back into the wheelchair.	☐ Yes ☐ Yes with difficulty ☐ No	☐ Fully ☐ Somewhat ☐ Not at all	☐ Always ☐ Sometimes ☐ Never	☐ Yes ☐ No
16	Opening a hinged door, moving the wheelchair through it and closing it behind you, then coming back the other way.	☐ Yes ☐ Yes with difficulty ☐ No	☐ Fully ☐ Somewhat ☐ Not at all	☐ Always ☐ Sometimes ☐ Never	☐ Yes ☐ No

Continued

		Questions (Pick only one answer for each question)			
#	Skill description	Can you do it?	How confident are you?	How often do you do it?	Is this a training goal?
17	Moving the wheelchair over a longer distance, for example, on a smooth surface about the length of a sport field.	☐ Yes ☐ Yes with difficulty ☐ No	☐ Fully ☐ Somewhat ☐ Not at all	☐ Always ☐ Sometimes ☐ Never	☐ Yes ☐ No
18	While moving the wheelchair, avoiding moving people who do not notice you.	☐ Yes ☐ Yes with difficulty ☐ No	☐ Fully ☐ Somewhat ☐ Not at all	☐ Always ☐ Sometimes ☐ Never	☐ Yes ☐ No
19	Moving the wheelchair up a slight incline, for example, a standard ramp (12 times longer than it is high).	☐ Yes ☐ Yes with difficulty ☐ No	☐ Fully ☐ Somewhat ☐ Not at all	☐ Always ☐ Sometimes ☐ Never	☐ Yes ☐ No
20	Moving the wheelchair down a slight incline.	☐ Yes ☐ Yes with difficulty ☐ No	☐ Fully ☐ Somewhat ☐ Not at all	☐ Always ☐ Sometimes ☐ Never	☐ Yes ☐ No
21	Moving the wheelchair up a steep incline (about twice as steep as a standard ramp).	☐ Yes ☐ Yes with difficulty ☐ No	☐ Fully ☐ Somewhat ☐ Not at all	☐ Always ☐ Sometimes ☐ Never	☐ Yes ☐ No
22	Moving the wheelchair down a steep incline.	☐ Yes ☐ Yes with difficulty ☐ No	☐ Fully ☐ Somewhat ☐ Not at all	☐ Always ☐ Sometimes ☐ Never	☐ Yes ☐ No
23	Moving the wheelchair across a slight side-slope, for example, when crossing a driveway.	☐ Yes ☐ Yes with difficulty ☐ No	☐ Fully ☐ Somewhat ☐ Not at all	☐ Always ☐ Sometimes ☐ Never	☐ Yes ☐ No
24	Moving the wheelchair a short distance across a soft surface, for example, gravel.	☐ Yes ☐ Yes with difficulty ☐ No	☐ Fully ☐ Somewhat ☐ Not at all	☐ Always ☐ Sometimes ☐ Never	☐ Yes ☐ No
25	Getting the wheelchair over an obstacle that sticks up above the surface, for example, a door threshold.	☐ Yes ☐ Yes with difficulty ☐ No	☐ Fully ☐ Somewhat ☐ Not at all	☐ Always ☐ Sometimes ☐ Never	☐ Yes ☐ No
26	Getting the wheelchair over a gap, for example, a rut in the road that is too big to simply roll over.	☐ Yes ☐ Yes with difficulty ☐ No	☐ Fully ☐ Somewhat ☐ Not at all	☐ Always ☐ Sometimes ☐ Never	☐ Yes ☐ No

Continued

#	Skill description	Questions (Pick only one answer for each question)			
		Can you do it?	How confident are you?	How often do you do it?	Is this a training goal?
27	Getting the wheelchair up a low curb, for example, when entering a building.	☐ Yes ☐ Yes with difficulty ☐ No	☐ Fully ☐ Somewhat ☐ Not at all	☐ Always ☐ Sometimes ☐ Never	☐ Yes ☐ No
28	Getting the wheelchair down from a low curb.	☐ Yes ☐ Yes with difficulty ☐ No	☐ Fully ☐ Somewhat ☐ Not at all	☐ Always ☐ Sometimes ☐ Never	☐ Yes ☐ No
29	Getting the wheelchair user up from the ground into the wheelchair, for example, after a fall.	☐ Yes ☐ Yes with difficulty ☐ No	☐ Fully ☐ Somewhat ☐ Not at all	☐ Always ☐ Sometimes ☐ Never	☐ Yes ☐ No

If you have any general comments about the questions that you have answered above, please record them in the space available below.

```

```

If you have any training goals that you have not already mentioned, please record them in the space available below.

```

```

A short report form will be created from the answers that you have given. If you would like a copy of the report form for yourself or someone else, please record in the space available below the name and address of the person to whom the report should be sent.

```

```

This is the end of the questionnaire. Thank you for completing it.

Wheelchair Skills Test Questionnaire (WST-Q), Version 4.3 for Scooters Operated by Their Users

Question	Answer
Name of the scooter user:	
Date questionnaire completed (month, day, year):	
Did you complete the questionnaire yourself?	☐ Yes ☐ No
If you had help, what is the name of the person who helped you?	
If you had help, what is the relationship between you and the person who helped you?	☐ Family member ☐ Friend ☐ Caregiver ☐ Other person

Introduction to the questionnaire

- Copies of this questionnaire can be downloaded from www.wheelchair-skillsprogram.ca/eng/wstq.php.
- More details about the questionnaire can be found earlier in this book or in the WSP Manual.
- In this questionnaire, you will be asked questions about different skills that you might do in your scooter. These skills range from ones that are more basic at the beginning to those that are more advanced at the end.
- There are no "right" or "wrong" answers. The purpose of the questionnaire is simply to help us understand how you use your scooter.
- It will probably take about 10 minutes to complete the questionnaire, but please take as much time as you need.
- If you have more than one scooter, the questions are about the scooter that you use most often.
- If you have any comments, you will be able to record them at the end of the questionnaire.
- For each specific skill, you will be asked up to four questions. The questions and the possible answers are shown below.
- For each skill, you should answer the following question:

Question: "Can you do it?"	
Possible answers	**What this means**
Yes	I can safely do the skill, by myself, without any difficulty.
Yes with difficulty	Yes, but not as well as I would like.
No	I have never done the skill or I do not feel that I could do it right now.
Not possible with this wheelchair	The scooter does not have the parts to allow this skill. (This option is only presented for skills where such a score is a possibility.)

- If one of the purposes of this questionnaire is to assess how confident you are in performing the skill, you should also answer the following question for each skill:

Confidence Question: "How confident are you?"	
Possible answers	**What this means**
Fully	As of now, I am fully confident that I can do this skill safely and consistently.
Somewhat	As of now, I am somewhat confident that I can do this skill safely and consistently.
Not at all	As of now, I am not at all confident that I can do this skill safely and consistently.

- If one of the purposes of this questionnaire is to assess how often you do the skill, you should also answer the following question for each skill:

Question: "How often do you do it?"	
Possible answers	**What this means**
Always	Whenever I need or want to do so.
Sometimes	Sometimes when I need or want to, sometimes not.
Never	Never or less often than once a year.

- If one of the purposes of this questionnaire is to identify goals for training, you should also answer the following question about each skill:

Question: "Is this a training goal?"	
Possible answers	**What this means**
Yes	I am interested in receiving training for this skill.
No	I am not interested in receiving training for this skill.

- If you have training goals that you can think of now, please record them in the space available below. You will have a chance to identify other goals later.

- Please read the questions about specific skills that begin on the next page. For each skill, record the answers in the spaces provided.

Questions on Specific Skills

#	Skill description	Questions (Pick only one answer for each question)			
		Can you do it?	How confident are you?	How often do you do it?	Is this a training goal?
1	Moving the controller away and back again.	☐ Yes ☐ Yes with difficulty ☐ No ☐ Not possible with this wheelchair	☐ Fully ☐ Somewhat ☐ Not at all	☐ Always ☐ Sometimes ☐ Never	☐ Yes ☐ No
2	Turning the power for the scooter on and off.	☐ Yes ☐ Yes with difficulty ☐ No	☐ Fully ☐ Somewhat ☐ Not at all	☐ Always ☐ Sometimes ☐ Never	☐ Yes ☐ No
3	Changing the settings and speeds for the scooter.	☐ Yes ☐ Yes with difficulty ☐ No ☐ Not possible with this wheelchair	☐ Fully ☐ Somewhat ☐ Not at all	☐ Always ☐ Sometimes ☐ Never	☐ Yes ☐ No
4	Disengaging the motors of the scooter, so that it can be pushed without power, and then engaging the motors again.	☐ Yes ☐ Yes with difficulty ☐ No	☐ Fully ☐ Somewhat ☐ Not at all	☐ Always ☐ Sometimes ☐ Never	☐ Yes ☐ No
5	Charging the battery for the scooter.	☐ Yes ☐ Yes with difficulty ☐ No	☐ Fully ☐ Somewhat ☐ Not at all	☐ Always ☐ Sometimes ☐ Never	☐ Yes ☐ No
6	Moving the scooter straight forward for a short distance, for example, along a short hallway.	☐ Yes ☐ Yes with difficulty ☐ No	☐ Fully ☐ Somewhat ☐ Not at all	☐ Always ☐ Sometimes ☐ Never	☐ Yes ☐ No
7	Moving the scooter straight backward for a short distance, for example, to back away from a table.	☐ Yes ☐ Yes with difficulty ☐ No	☐ Fully ☐ Somewhat ☐ Not at all	☐ Always ☐ Sometimes ☐ Never	☐ Yes ☐ No
8	Turning the scooter around in a small space so that it is facing in the opposite direction.	☐ Yes ☐ Yes with difficulty ☐ No	☐ Fully ☐ Somewhat ☐ Not at all	☐ Always ☐ Sometimes ☐ Never	☐ Yes ☐ No
9	Turning the scooter around a corner while moving forward.	☐ Yes ☐ Yes with difficulty ☐ No	☐ Fully ☐ Somewhat ☐ Not at all	☐ Always ☐ Sometimes ☐ Never	☐ Yes ☐ No

Continued

#	Skill description	Can you do it?	How confident are you?	How often do you do it?	Is this a training goal?
		Questions (Pick only one answer for each question)			
10	Turning the scooter around a corner while moving backward.	☐ Yes ☐ Yes with difficulty ☐ No	☐ Fully ☐ Somewhat ☐ Not at all	☐ Always ☐ Sometimes ☐ Never	☐ Yes ☐ No
11	Moving the scooter sideways in a small space, for example, to get the side of the scooter next to a kitchen counter, and then back to where you started.	☐ Yes ☐ Yes with difficulty ☐ No	☐ Fully ☐ Somewhat ☐ Not at all	☐ Always ☐ Sometimes ☐ Never	☐ Yes ☐ No
12	Moving the scooter to reach up for something overhead, for example, a high elevator button.	☐ Yes ☐ Yes with difficulty ☐ No	☐ Fully ☐ Somewhat ☐ Not at all	☐ Always ☐ Sometimes ☐ Never	☐ Yes ☐ No
13	Moving the scooter to pick up a small object, for example, a paperback book, from the floor in front of you.	☐ Yes ☐ Yes with difficulty ☐ No	☐ Fully ☐ Somewhat ☐ Not at all	☐ Always ☐ Sometimes ☐ Never	☐ Yes ☐ No
14	Operating all of the positioning options of the scooter (e.g., tilting the seat).	☐ Yes ☐ Yes with difficulty ☐ No ☐ Not possible with this wheelchair	☐ Fully ☐ Somewhat ☐ Not at all	☐ Always ☐ Sometimes ☐ Never	☐ Yes ☐ No
15	Transferring from the scooter to a bench that is about the same height as the scooter and then getting back into the wheelchair.	☐ Yes ☐ Yes with difficulty ☐ No	☐ Fully ☐ Somewhat ☐ Not at all	☐ Always ☐ Sometimes ☐ Never	☐ Yes ☐ No
16	Opening a hinged door, moving the scooter through it and closing it behind you, then coming back the other way.	☐ Yes ☐ Yes with difficulty ☐ No	☐ Fully ☐ Somewhat ☐ Not at all	☐ Always ☐ Sometimes ☐ Never	☐ Yes ☐ No
17	Moving the scooter over a longer distance, for example, on a smooth surface about the length of a sport field.	☐ Yes ☐ Yes with difficulty ☐ No	☐ Fully ☐ Somewhat ☐ Not at all	☐ Always ☐ Sometimes ☐ Never	☐ Yes ☐ No

Continued

#	Skill description	Questions (Pick only one answer for each question)			
		Can you do it?	How confident are you?	How often do you do it?	Is this a training goal?
18	While moving the scooter, avoiding moving people who do not notice you.	☐ Yes ☐ Yes with difficulty ☐ No	☐ Fully ☐ Somewhat ☐ Not at all	☐ Always ☐ Sometimes ☐ Never	☐ Yes ☐ No
19	Moving the scooter up a slight incline, for example, a standard ramp (12 times longer than it is high).	☐ Yes ☐ Yes with difficulty ☐ No	☐ Fully ☐ Somewhat ☐ Not at all	☐ Always ☐ Sometimes ☐ Never	☐ Yes ☐ No
20	Moving the scooter down a slight incline.	☐ Yes ☐ Yes with difficulty ☐ No	☐ Fully ☐ Somewhat ☐ Not at all	☐ Always ☐ Sometimes ☐ Never	☐ Yes ☐ No
21	Moving the scooter up a steep incline (about twice as steep as a standard ramp).	☐ Yes ☐ Yes with difficulty ☐ No	☐ Fully ☐ Somewhat ☐ Not at all	☐ Always ☐ Sometimes ☐ Never	☐ Yes ☐ No
22	Moving the scooter down a steep incline.	☐ Yes ☐ Yes with difficulty ☐ No	☐ Fully ☐ Somewhat ☐ Not at all	☐ Always ☐ Sometimes ☐ Never	☐ Yes ☐ No
23	Moving the scooter across a slight side-slope, for example, when crossing a driveway.	☐ Yes ☐ Yes with difficulty ☐ No	☐ Fully ☐ Somewhat ☐ Not at all	☐ Always ☐ Sometimes ☐ Never	☐ Yes ☐ No
24	Moving the scooter a short distance across a soft surface, for example, gravel.	☐ Yes ☐ Yes with difficulty ☐ No	☐ Fully ☐ Somewhat ☐ Not at all	☐ Always ☐ Sometimes ☐ Never	☐ Yes ☐ No
25	Getting the scooter over an obstacle that sticks up above the surface, for example, a door threshold.	☐ Yes ☐ Yes with difficulty ☐ No	☐ Fully ☐ Somewhat ☐ Not at all	☐ Always ☐ Sometimes ☐ Never	☐ Yes ☐ No
26	Getting the scooter over a gap, for example, a rut in the road that is too big to simply roll over.	☐ Yes ☐ Yes with difficulty ☐ No	☐ Fully ☐ Somewhat ☐ Not at all	☐ Daily ☐ Weekly ☐ Monthly ☐ Yearly ☐ Never	☐ Yes ☐ No
27	Getting the scooter up a low curb, for example, when entering a building.	☐ Yes ☐ Yes with difficulty ☐ No	☐ Fully ☐ Somewhat ☐ Not at all	☐ Daily ☐ Weekly ☐ Monthly ☐ Yearly ☐ Never	☐ Yes ☐ No

Continued

#	Skill description	Can you do it?	How confident are you?	How often do you do it?	Is this a training goal?
			Questions (Pick only one answer for each question)		
28	Getting the scooter down from a low curb.	☐ Yes ☐ Yes with difficulty ☐ No	☐ Fully ☐ Somewhat ☐ Not at all	☐ Daily ☐ Weekly ☐ Monthly ☐ Yearly ☐ Never	☐ Yes ☐ No
29	Getting up from the ground into the scooter, for example, after a fall.	☐ Yes ☐ Yes with difficulty ☐ No	☐ Fully ☐ Somewhat ☐ Not at all	☐ Daily ☐ Weekly ☐ Monthly ☐ Yearly ☐ Never	☐ Yes ☐ No

If you have any general comments about the questions that you have answered above, please record them in the space available below.

```



```

If you have any training goals that you have not already mentioned, please record them in the space available below.

```



```

A short report form will be created from the answers that you have given. If you would like a copy of the report form for yourself or someone else, please record in the space available below the name and address of the person to whom the report should be sent.

```



```

This is the end of the questionnaire. Thank you for completing it.

Appendix 5: WST-Q report forms

Wheelchair Skills Test Questionnaire (WST-Q), Version 4.3
Manual Wheelchairs Operated by Wheelchair Users

Name of the wheelchair user: _____ Date:_____

Person completing questionnaire (if not user): _____

Relationship between the wheelchair user and the person who helped him/her: _____

#	Individual Skill	Capacity (0–2)	Confidence (0–2)	Performance (0–2)	Composite (0–6)
1	Rolls forward short distance				
2	Rolls backward short distance				
3	Turns in place				
4	Turns while moving forward				
5	Turns while moving backward				
6	Maneuvers sideways				
7	Reaches high object				
8	Picks object from floor				
9	Relieves weight from buttocks				
10	Operates body positioning options				
11	Level transfer				
12	Folds and unfolds wheelchair				
13	Gets through hinged door				
14	Rolls longer distance				
15	Avoids moving obstacles				
16	Ascends slight incline				
17	Descends slight incline				
18	Ascends steep incline				
19	Descends steep incline				
20	Rolls across side-slope				
21	Rolls on soft surface				
22	Gets over threshold				
23	Gets over gap				
24	Ascends low curb				
25	Descends low curb				
26	Ascends high curb				
27	Descends high curb				
28	Performs stationary wheelie				
29	Turns in place in wheelie position				
30	Descends high curb in wheelie position				
31	Descends steep incline in wheelie position				
32	Gets from ground into wheelchair				
33	Ascends stairs				
34	Descends stairs				
	Total scores (%):				

General comments:

Training goals described by the wheelchair user:

Name and address of any person(s) to whom the test subject would like a copy of the report to be sent.

Details about the WST-Q can be found at: www.wheelchairskillsprogram.ca/eng/manual.php/.

Appendix 5 _____

Wheelchair Skills Test Questionnaire (WST-Q), Version 4.3
Manual Wheelchairs Operated by Caregivers

Name of the caregiver: _____ Date:_____

Person completing questionnaire (if not user): _____

Relationship between the wheelchair user and the person who helped him/her: _____

#	Individual Skill	Capacity (0–2)	Confidence (0–2)	Performance (0–2)	Composite (0–6)
1	Rolls forward short distance				
2	Rolls backward short distance				
3	Turns in place				
4	Turns while moving forward				
5	Turns while moving backward				
6	Maneuvers sideways				
7	Picks object from floor				
8	Relieves weight from buttocks				
9	Operates body positioning options				
10	Level transfer				
11	Folds and unfolds wheelchair				
12	Gets through hinged door				
13	Rolls longer distance				
14	Avoids moving obstacles				
15	Ascends slight incline				
16	Descends slight incline				
17	Ascends steep incline				
18	Descends steep incline				
19	Rolls across side-slope				
20	Rolls on soft surface				
21	Gets over threshold				
22	Gets over gap				
23	Ascends low curb				
24	Descends low curb				
25	Ascends high curb				
26	Descends high curb				
27	Performs stationary wheelie				
28	Turns in place in wheelie position				
29	Descends high curb in wheelie position				
30	Descends steep incline in wheelie position				
31	Gets from ground into wheelchair				
32	Ascends stairs				
33	Descends stairs				
	Total scores (%):				

General comments:

Training goals described by the caregiver:

Name and address of any person(s) to whom the test subject would like a copy of the report to be sent.

Details about the WST-Q can be found at: www.wheelchairskillsprogram.ca/eng/manual.php/.

Wheelchair Skills Test Questionnaire (WST-Q), Version 4.3
Powered Wheelchairs Operated by Wheelchair Users

Name of the wheelchair user: _____ Date:_____

Person completing questionnaire (if not user): _____

Relationship between the wheelchair user and the person who helped him/her: _____

#	Individual Skill	Capacity (0–2)	Confidence (0–2)	Performance (0–2)	Composite (0–6)
1	Moves controller away and back				
2	Turns power on and off				
3	Selects drive modes and speeds				
4	Disengages and engages motors				
5	Operates battery charger				
6	Rolls forward short distance				
7	Rolls backward short distance				
8	Turns in place				
9	Turns while moving forward				
10	Turns while moving backward				
11	Maneuvers sideways				
12	Reaches high object				
13	Picks object from floor				
14	Relieves weight from buttocks				
15	Operates body positioning options				
16	Level transfer				
17	Gets through hinged door				
18	Rolls longer distance				
19	Avoids moving obstacles				
20	Ascends slight incline				
21	Descends slight incline				
22	Ascends steep incline				
23	Descends steep incline				
24	Rolls across side-slope				
25	Rolls on soft surface				
26	Gets over threshold				
27	Gets over gap				
28	Ascends low curb				
29	Descends low curb				
30	Gets from ground into wheelchair				
	Total scores (%):				

General comments:

Training goals described by the wheelchair user:

Name and address of any person(s) to whom the test subject would like a copy of the report to be sent.

Details about the WST-Q can be found at: www.wheelchairskillsprogram.ca/eng/manual.php/.

Appendix 5 _____

Wheelchair Skills Test Questionnaire (WST-Q), Version 4.3

Powered Wheelchairs Operated by Caregivers

Name of the caregiver: _____ Date:_____

Person completing questionnaire (if not user): _____

Relationship between the wheelchair user and the person who helped him/her: _____

#	Individual Skill	Capacity (0–2)	Confidence (0–2)	Performance (0–2)	Composite (0–6)
1	Moves controller away and back				
2	Turns power on and off				
3	Selects drive modes and speeds				
4	Disengages and engages motors				
5	Operates battery charger				
6	Rolls forward short distance				
7	Rolls backward short distance				
8	Turns in place				
9	Turns while moving forward				
10	Turns while moving backward				
11	Maneuvers sideways				
12	Picks object from floor				
13	Relieves weight from buttocks				
14	Operates body positioning options				
15	Level transfer				
16	Gets through hinged door				
17	Rolls longer distance				
18	Avoids moving obstacles				
19	Ascends slight incline				
20	Descends slight incline				
21	Ascends steep incline				
22	Descends steep incline				
23	Rolls across side-slope				
24	Rolls on soft surface				
25	Gets over threshold				
26	Gets over gap				
27	Ascends low curb				
28	Descends low curb				
29	Gets from ground into wheelchair				
	Total scores (%):				

General comments:

Training goals described by the caregiver:

Name and address of any person(s) to whom the test subject would like a copy of the report to be sent.

Details about the WST-Q can be found at: www.wheelchairskillsprogram.ca/eng/manual.php/.

Wheelchair Skills Test Questionnaire (WST-Q), Version 4.3

Scooters Operated by Their Users

Name of the scooter user: _____ Date:_____

Person completing questionnaire (if not user): _____

Relationship between the wheelchair user and the person who helped him/her: _____

#	Individual Skill	Capacity (0–2)	Confidence (0–2)	Performance (0–2)	Composite (0–6)
1	Moves controller away and back				
2	Turns power on and off				
3	Selects drive modes and speeds				
4	Disengages and engages motors				
5	Operates battery charger				
6	Rolls forward short distance				
7	Rolls backward short distance				
8	Turns in place				
9	Turns while moving forward				
10	Turns while moving backward				
11	Maneuvers sideways				
12	Reaches high object				
13	Picks object from floor				
14	Operates body positioning options				
15	Level transfer				
16	Gets through hinged door				
17	Rolls longer distance				
18	Avoids moving obstacles				
19	Ascends slight incline				
20	Descends slight incline				
21	Ascends steep incline				
22	Descends steep incline				
23	Rolls across side-slope				
24	Rolls on soft surface				
25	Gets over threshold				
26	Gets over gap				
27	Ascends low curb				
28	Descends low curb				
29	Gets from ground into scooter				
	Total scores (%):				

General comments:

Training goals described by the scooter user:

Name and address of any person(s) to whom the test subject would like a copy of the report to be sent.

Details about the WST-Q can be found at: www.wheelchairskillsprogram.ca/eng/ manual.php/.

Index

Printed in the United States
by Baker & Taylor Publisher Services